Y0-CCV-592

Hydraulic Valves and Controls

FLUID POWER AND CONTROL

A Series of Textbooks and Reference Books

1. Hydraulic Pumps and Motors: Selection and Application for Hydraulic Power Control Systems, *by Raymond P. Lambeck*
2. Designing Pneumatic Control Circuits: Efficient Techniques for Practical Application, *by Bruce E. McCord*
3. Fluid Power Troubleshooting, *by Anton H. Hehn*
4. Hydraulic Valves and Controls: Selection and Application, *by John J. Pippenger*

Other Volumes in Preparation

Hydraulic Valves and Controls

SELECTION AND APPLICATION

John J. Pippenger

Educational Coordinator
Fluid Power Educational Foundation
Milwaukee, Wisconsin

MARCEL DEKKER, INC. NEW YORK AND BASEL

Library of Congress Cataloging in Publication Data

Pippenger, John J.
Hydraulic valves and controls.

(Fluid power and control ; 4)
Includes index.
1. Hydraulic control. 2. Valves. 3. Fluid power technology. I. Title. II. Series.
TJ843.P47 1984 621.2 83-20980
ISBN 0-8247-7087-0

MARCEL DEKKER, INC.
270 Madison Avenue, New York, New York 10016

Current printing (last digit) :
10 9 8 7 6 5 4 3 2 1

PRINTED IN THE UNITED STATES OF AMERICA

Foreword

The author of this volume is exceptionally well qualified to discuss the selection and applications of hydraulic valves and controls. He has worked in that field for over 40 years with respected fluid power companies such as Double A, Racine, and HPI/Nichols, each of which has been responsible for the development of novel pump or control valve technology.

His thoughtful involvement in the activities of the JIC (Joint Industry Conference), in the development of the original JIC Hydraulic Standards and subsequent series of updates, his participation in the American National Standards Institute B93 (Fluid Power) Committee and participation in the National Fluid Power Association Standards activities have given him an exceptional overview of hydraulic system design practices in the USA.

Because of his extensive travels in Europe and the Orient, he has had the opportunity to study applications of hydraulic valves and controls as practiced by the world's major manufacturers and users of hydraulics. Further, as a direct result of his activity in the National Conference on Industrial Hydraulics, now known as the National Conference of Fluid Power, and in the activities of ISO Technical Committee 131, which deals with fluid power components and applications, he has gained considerable insight into future developments which will soon impact the hydraulic industry.

In his preface, John Pippenger states that technological developments have resulted in a high degree of sophistication being introduced to fluid power systems. In the current "electronic revolution," electronic controls are providing designers of hydraulic controls the tools to economically achieve hydraulic system control that until now was only a dream. Circuit and machine designers who understand basic hydraulic control and pump technology will be able to readily utilize the application of electronics.

Within the past decade, electronically actuated hydraulic control systems have evolved from combinations of fixed displacement pumps and "bang-bang" solenoid driven two- or three-position controls to proportional electrically controlled variable displacement pumps and a broad selection of pressure, flow, and direction control valves with an almost infinite range of control.

Circuit design has progressed from sequential step programming accomplished manually with visual feedback by a human operator. Circuit control by a programmed series of steps where fluid controls responded to the sequential commands of a programmable controller or other step type control has proved to be only an interim circuit design solution. Today's technology allows—if not invites—the circuit designer to use proportional pressure and flow control valves in conjunction with almost infinitely positionable directional control valves, all controlled by electronic logic devices.

Given the current availability of sensors at a reasonable cost, especially position and pressure sensors, the design of closed loop electronically controlled hydraulic systems should be considered for many applications. Many designers will be pleasantly surprised at the performance and overall economics of such an approach.

The author goes well beyond mere description of typical hydraulic hardware. He provides the reader with a solid background of typical applications for each type of control discussed with sufficient commentary to assist the serious circuit designer to develop good selection practices and to structure hydraulic circuits which will be reliable, efficient, quiet, and productive.

This volume is a basic must for any machine designer striving to produce a better hydraulically powered machine.

Z. J. Lansky

Preface

Energy conversion is basic to our global economy. Fluid power systems have become a significant part of the power transmission portion of the energy conversion systems.

The title *fluid power* was coined to identify hydraulics (liquid as the fluid medium), pneumatics (gas as the fluid medium), and combinations of hydraulics and pneumatics in the power transmission system.

Within the last half-century, a high degree of sophistication has been introduced into virtually every phase of energy conservation and energy conversion. Fluid power systems have been no exception.

An objective of this book is to acquaint the student, practicing engineer, and/or technician with the valves and controls available to structure the hydraulic circuits that are widely used in the energy conversion chain.

The hydraulic circuits we will consider generally consist of a pump to move and pressurize a liquid, suitable controls for pressure level, direction, and rate of flow, and a hydraulic cylinder to provide linear force and motion and/or a hydraulic motor to provide rotary force and motion.

Our interest is primarily in hydraulic systems employing a contained liquid as the power transmitting agent as compared to a pneumatic system that employs a contained gas as the power transmitting agent. Further, our interest in the flow of the liquid is strictly as a power transmitting agent in a hydrostatic system as compared with hydrodynamics wherein the movement of the fluid is the significant factor.

Support from industry is vital in the preparation of technical works such as this presentation. Major contributors of illustrations and editing services include Sun Hydraulics Corporation, Continental Hydraulics, Moog Inc.,

Eaton Corporation Fluid Power Operations, HPI/Nichols, Rexnord Fluid Power Division,* Kepner Products Company, McGraw-Hill Book Co., Barksdale Division Transamerica Delaval, Hydreco Division Union Signal Corporation, Sperry Vickers, Double A Products Division Brown & Sharp, The Rexroth Corporation, and the Parker Hannifin Corporation.

To recognize individuals who have provided guidance and counsel would result in a listing that might be best described as a "whose-who" in the fluid power industry. My sincere thanks go to these individuals and corporations that have made this book possible.

John J. Pippenger

*Now a division of Dana Corp.

Contents

Hydraulic Valves and Controls

1 Characteristics of Hydraulic Circuits

1.1. INTRODUCTION: HYDRAULIC SYSTEMS

Transmission of power with a contained liquid as an essential link in a transmission system can be accomplished by two separate and distinct systems.

1.1.1. Hydrodynamic Systems

Hydrodynamic systems depend upon the inertia of the moving fluid to accomplish the desired power transmission function. These systems are called hydrodynamic because of the energy transfer pattern. The hydraulic coupling in an automotive-type automatic transmission provides a good example of a hydrodynamic power transmission system.

1.1.2. Hydrostatic Systems

Liquid contained within an enclosed conductor is moved and pressurized by a positive-displacement pumping mechanism in a hydrostatic-type system. The energy implanted by the pump mechanism on the liquid is then available to move a linear hydraulic cylinder for push or pull action or to a rotary hydraulic motor for rotary force and motion. Pumps can be structured to accomplish the desired liquid movement with several different basic design characteristics.

1.2. PUMP CHARACTERISTICS

1.2.1. Centrifugal

Certain nonpositive displacement pumps do find use in hydrostatic type systems. A centrifugal-type pump can move and pressurize liquid by use of a series of paddles in an enclosed passage that is located in an impeller in the pump housing. The housing is designed with appropriate passages to accept the liquid at the center of the drive and paddle assembly and move it to the outer periphery of this blade-like impeller with an outlet port at the outer circumference of the housing (Fig. 1.1). A limited increase in pressure is accomplished with a device of this type. To increase potential pressure development it is common to design and build centrifugal pumps with multiple sections or stages connected in series to provide the desired output pressure capabilities.

Centrifugal pumps are often used to move and pressurize large quantities of liquid for filling and hydrostatically testing large pressure vessels or pipe. Smaller units are also used to move and pressure liquid from atmospheric pressure to values somewhat higher thereby supplying the input to other pump types and improving the fill characteristics of the higher pressure-type pumps. Centrifugal pumps are commonly used in passenger and freight elevator systems.

1.2.2. Propeller

Propeller-type pumps serve a similar boost function for higher pressure pumps. The propeller-type pump (Fig. 1.2), as the name implies, consists of a paddle-like structure which moves the fluid in a linear path through the conductor at a rate appropriate to the rotary speed, blade area, and characteristics of the fluid. Centrifugal and propeller-type pumps are nonpositive displacement pumps. Flow is dependent on rotative speed and resistance to movement at the outlet port.

1.2.3. Piston, Vane, Gear, and Screw-Type Pumps

Piston, vane, gear, and screw-type pumps are positive displacement-type devices. A fixed quantity of liquid is moved at each revolution of the drive shaft (Fig. 1.3).

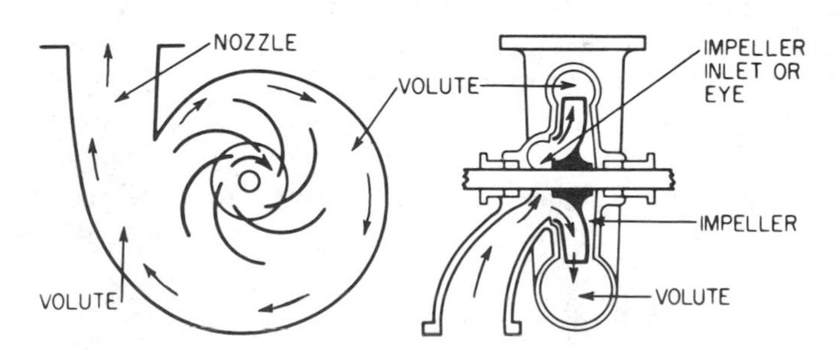

Figure 1.1 Fluid flow through a centrifugal pump. (From Pippenger, John J. and Hicks, Tyler G., *Industrial Hydraulics,* 3rd Edition, Gregg Division, McGraw-Hill Book Co., New York, New York, 1979.)

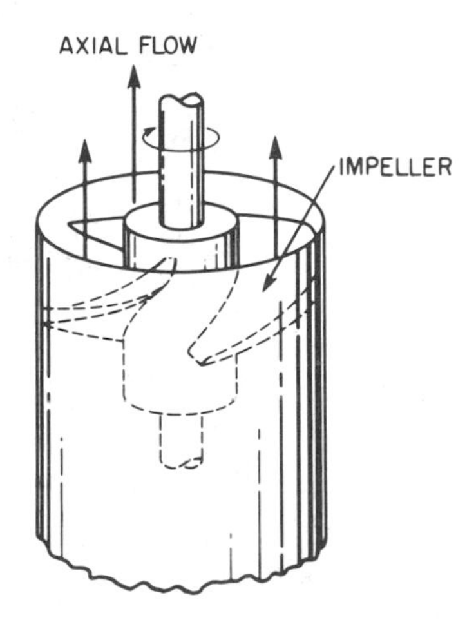

Figure 1.2 Propeller pump moves fluid in an axial flow. (From Pippenger, John J. and Hicks, Tyler G., *Industrial Hydraulics,* 3rd Edition, Gregg Division, McGraw-Hill Book Co., New York, New York, 1979.)

If the displacement mechanism cannot be altered, the pump is called a fixed-displacement-type. Most gear-type pumps are fixed displacement machines. A known quantity of liquid will be moved at each revolution of the drive shaft.

1.2.4. Efficiency

Leakage past the various moving pump mechanisms will affect the net amount of liquid that will be available at the outlet port. Some leakage is usually designed into the pump structure to lubricate and cool the working members. These leakages or predetermined clearance flows affect the efficiency of the pump and the overall power transmission function. Efficiency relates to the total energy recovered in movement of the output members in the machine circuit. Some energy may be sacrificed to increase the life span of a component.

1.2.5. Variable Displacement

Piston and vane-type pump structures can be designed and manufactured with facilities to change the displacement at the desire of the machine operator or designer. This function can be manual or automatic. Such a pump is called a variable displacement device. A predetermined quantity of fluid will be moved by a variable displacement pump according to the adjustment of the displacement structure. If the displacement is reduced to zero discharge, flow will cease. Usually a small flow is provided for lubrication and cooling.

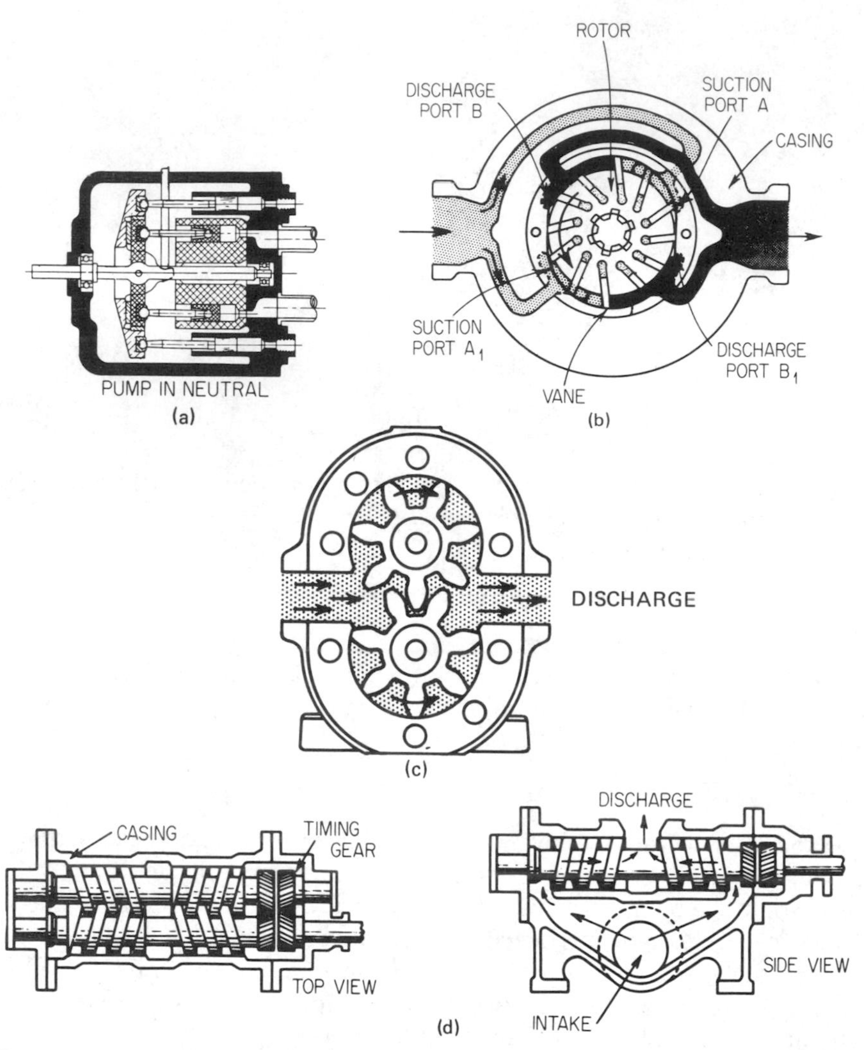

Figure 1.3 (a) Piston pump employs one or more pistons in a bore with suitable valving to separate suction from discharge. (Courtesy of Rexnord, Inc. Fluid Power Division, Racine, Wisconsin.*) (b) Vanes follow a contoured path to create a pumping pattern in a suitable housing. (Courtesy of Sperry Vickers, Troy, Michigan.) (c) Spur gear pump provides rugged displacement action for pumping function. (Courtesy of Rexnord, Inc. Fluid Power Division, Racine, Wisconsin.) (d) Screw-type pump provides quiet, dependable operation as pumping mechanism. (Courtesy of Transamerica Delavel, Inc., Los Angeles, California.)

*Now a division of Dana Corp.

1.3. CIRCUIT CHARACTERISTICS

1.3.1. Open Hydraulic Circuits

Fixed displacement pumps are usually used in *open hydraulic* circuits. An open hydraulic circuit contains at least one pump supplied with liquid from a tank or reservoir, usually at atmospheric pressure. Reservoirs can be sealed and pressurized to minimize entry of foreign matter or to assist movement of the fluid into the pump inlet. The discharge of the pump or pumps is directed through appropriate valves to the hydraulic cylinder or motor there by providing the desired linear or rotary force and motion. Fluid returned from the valves is directed to the reservoir. Variable displacement pumps are also used in open hydraulic circuits at the designers option.

1.3.2. Closed Hydraulic Circuits

A variable displacement pump can be structured to reverse flow through the two connecting ports so that the valving function may be integrated into the pumping structure.

The key element of the open hydraulic system is the tank or reservoir of significant size where the spent fluid is returned prior to recycling through the pump or pumps. A closed hydraulic circuit usually consists of one variable displacement pump which can pump liquid in and out of each port according to the position of the control element and one hydraulic motor whose inlet and outlet ports are connected to the two ports of the pump. The direction of rotation of the motor will be governed by the flow pattern from the pump as determined by the control mechanism.

The reservoir or tank assembly, which may be within the pump housing, is usually small in a closed hydraulic system.The function of this small tank assembly or reservoir is to accept lubrication and cooling flows and provide a supply for a small auxiliary charge pump that pressurizes the captive liquid in the lines between the pump and output hydraulic motor. The small auxiliary pump replenishes the flow in the primary power loop used for lubrication, cooling and, at times, for a control pressure source for the closed hydraulic circuit in the hydrostatic transmission.

Linear cylinders can be incorporated in a closed hydraulic system. If the cylinder is provided with a rod extension of equal size from each end, the displacement will be equal in each direction of movement and the circuit characteristics will be similar to a rotary hydraulic motor. If the rod extends through one end of the cylinder only, there will be a displacement difference equal to the displacement of the piston rod. A larger reservoir may be required in the closed hydraulic circuit to accommodate the rod displacement plus valves to accommodate the differences in fluid usage in each direction of movement.

1.3.3. Traction Drive Systems

Many traction drive systems for agricultural and construction machines employ a hydrostatic transmission which consists of a reversible outlet piston pump and a fixed displacement motor. The pump is connected to the prime mover, often a gasoline or diesel engine. The hydraulic motor is connected directly to the wheel drive or track drive assembly. The wheel motor may be integrated into the wheel assembly for space savings and construction economy.

The prime mover can operate at the most efficient speed and the output drive speed can be set by the displacement of the hydrostatic transmission pump at any value appropriate to the task to be accomplished. Use of a variable displacement motor permits additional speed/torque range capabilities.

1.3.4. Marine Applications

Hydrostatic transmissions find wide usage in marine installations. The closed hydraulic circuit is ideal for winches and crane drives because of the flexibility of control and the limited number of components exposed to this hostile environment. Ice and extremes of temperature can be tolerated by the hydrostatic transmission systems because the liquid is usually a petroleum or synthetic oil that is totally contained within the hydraulic circuit. Circuit modifications are provided to circulate warm liquid to avoid unnecessary resistance to flow because of high viscosity conditions associated with cold liquids.

1.4. HYDRAULIC FLUIDS

1.4.1. Petroleum Based Fluids, Synthetics

Most hydrostatic transmissions operate with petroleum oil as the power transmission agent. Some systems may be supplied with synthetic fluids for special service such as extreme cold or heat wherein petroleum could ignite or be a hazard in the event of external leakage.

1.4.2. Aqueous Based Fluids

Some large presses may be designed to be compatible with clean water or water and oil combinations. The purpose may be to reduce cost of the fluid where large quantities are involved or it may be to eliminate fire hazards. Various resistance to ignition can be calculated according to the needs of the specific applications.

Machines using oil and water fluids can be equipped with hydraulic components usually employed with petroleum oils if certain precautions are observed and duty is tempered to the lubrication values of the fluid in service.

1.5. SUMMARY

The first objective of this book is to acquaint the student with the controls employed in the hydraulic circuit for direction, pressure level, and rate of flow of the fluid. A second objective is to acquaint the student with the interrelationship and develop communication skills through circuit drawings, symbology, and terminology.

By acquiring these skills, the student will be prepared to design control circuits for hydraulic systems and understand existing systems for purposes of maintenance and/or repair.

2 Symbology

2.1. INTRODUCTION: GRAPHIC SYMBOLS

Graphic symbols provide a quick, convenient means of identifying fluid power components and their associated relationships in a hydraulic circuit. The fluid flow pattern is clearly identified for all working and rest conditions within the hydraulic circuit. The actuation and control of the machinery by the hydraulic system is easily identified.

2.1.1. Circuit Drawings

Circuit drawings are used to show the components involved and the interconnections between the components used to pressurize captive fluid, and control pressure level, direction, and rate of flow of fluid through the hydraulic power transmission system. The *component symbols* and circuit drawing system communicate information relative to direction of flow, type of control, and certain relationships of physical *position* of components. The symbol and circuit data does *not* show detailed construction or physical size of components or of the interconnecting lines. The symbols and circuit drawing system used to communicate information relative to the hydraulic power transmission system from the machine designer to the machine fabricator and then to the maintenance personnel has been developed through several chronological steps.

2.1.2. History

Early development of the system was under the guidance of engineering personnel associated with the automotive manufacturing sector. The first widely used system was published by the Joint Industry Council which later became the Joint Industry Conference. This early milestone in the development of a symbology system is identified as the JIC system. The American Standards Association (ASA), which is the predecessor of the American National Standards Institute (ANSI) sponsored committee actions to refine and expand the JIC system in cooperation with the American Society of Mechanical Engineers (ASME).

A parallel activity by the International Standards Organization (ISO) has further assisted in the refinement of the system to make it an acceptable worldwide communication system with minimal problems associated with basic language differences. The similarity between the early JIC communication system and the ANSI/ISO documents helps to identify components and fluid flow on older machine service drawings. Many industrial and public libraries are able to provide access to texts describing the earlier systems if needed.

2.2. LINE GRAPHICS

2.2.1. Interconnecting Lines

Four basic circuit lines are used to identify the basic fluid conducting lines. A solid line is used to show power input, pump suction, and major return to tank conductors. A dash line is used to identify pilot fluid conductors. The pilot fluid conductors are usually of small size compared to the major supply and return conductors. A hyphenated line is used to identify return to tank lines that are provided to drain various areas within a circuit component. Two or more components may be physically located in a common housing or manifolded together with suitable fasteners to make the assembly an integrated unit. An enclosure line [Fig. 2.1(a)], usually in the form of a rectangle, identifies the quantity and functions of the components in the assembly.

2.2.2. Lines Crossing

A loop in one line as shown in Fig. 2.1(b) at the point of line crossing provides positive identification of the crossing function and precludes a connection identification. The earlier circuit drawings may not have the loops to indicate the crossing line. Particular care may be necessary to follow the circuit pattern on those drawings that do not use the loop crossing identification.

2.2.3. Lines Connecting

A dot equal to approximately five line widths [Fig. 2.1(c)] at the juncture of two or more lines indicate physical interconnection. Some drawings may be encountered that do not use the dot to show this interconnection. A line may

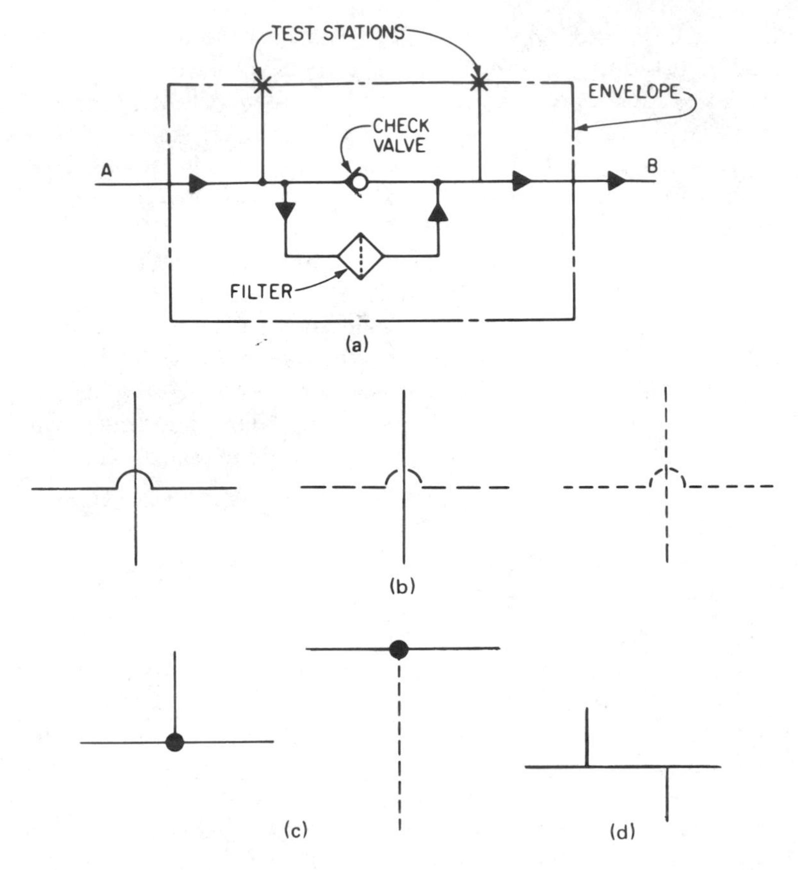

Figure 2.1 (a) Enclosure envelope. (b) Lines crossing. (c) Lines connecting with a dot. (d) Lines abutting.

abut another line at a right angle to indicate a connection [Fig. 2.1(d)]. The dot is the preferred identification of the point of juncture.

2.2.4. Lines to Tank or Reservoir

A *U-shaped* symbol [Fig. 2.2(a)] is used to indicate a tank connection. Termination of the line from the component at a point midway between the upright legs of the U-shaped symbol at a point from the bottom approximately equal to the length of the upright leg [Fig. 2.2(c)] indicates that the physical connection of the line must be *above* the highest anticipated fluid level. Termination of the line from the component at a point midway between the upright legs of the U-shaped symbol *touching* the bottom [Fig. 2.2(b)] indicates that the line must be *below* the lowest anticipated fluid level.

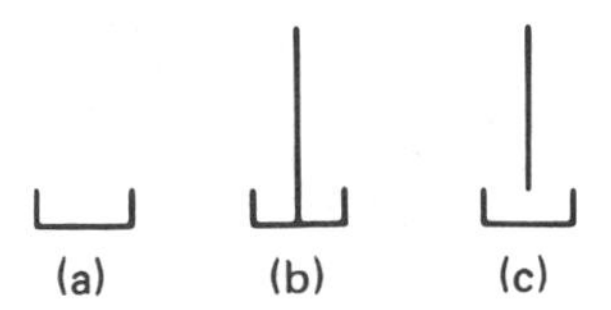

Figure 2.2 (a) Tank symbol. (b) Line terminating below fluid level. (c) Line terminating above fluid level.

Figure 2.3 Flexible line.

Figure 2.4 (a) Quick disconnect. (b) Quick disconnect with check valve.

A significant potential for circuit malfunction can be anticipated if the termination location of the return-to-tank rules are violated by careless drawing procedures. Some lines are obvious. A pump intake line above the fluid level may never be encountered. A drain line from a control component may not be as obvious. Certain drain lines may function best if they are below the fluid level to avoid aeration of the fluid or ingestion of free air into the component. Many drain lines need to terminate above the fluid level to avoid a pumping action which may adversely affect the component function.

2.2.5. Flexible Lines

The loop-type symbol (Fig. 2.3) indicates a flexible line.

2.2.6. Quick Connectors

A junction of two lines with a connector that can be easily connected or disconnected, usually manually without tools, is shown in Fig. 2.4(a). The inclusion of a check-valve-type device to prevent loss of fluid when the connector is separated is indicated by the symbol as shown in Fig. 2.4(b). Usually the quick connectors are installed between a rigid and a flexible line. The fluid retaining device may be in one or both lines. Quick connectors are used in such applications as cylinder lines to agricultural implements and fluid lines to contractors construction tools.

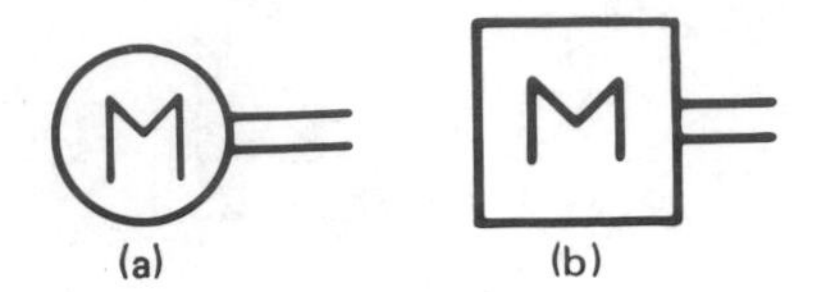

Figure 2.5 (a) Electric motor. (b) Heat engine (Internal Combustion Engine).

2.3. FLUID MOVEMENT AND PRESSURIZATION

2.3.1. Prime Mover and Pump Symbols

An electric motor is identified by a symbol as shown in Fig. 2.5(a). A heat engine is identified by a symbol as shown in Fig. 2.5(b).

The basic symbol for a pump or hydraulic motor is a circle [Fig. 2.6(a)]. An equilateral triangle shows direction of flow through the pump. The solid equilateral triangle [Fig. 2.6(b)] indicates that the pump is used to move and pressurize a liquid. A hollow equilateral triangle [Fig. 2.6(c)] indicates that gas or air is to be moved and pressurized by the device. Hydrualic pumps are always identified by a round basic symbol with at least one solid equilateral triangle.

2.3.2. Mechanical Connections Between Prime Mover and Pump

Figure 2.7(a) shows the pump and prime mover connected. The circular arrow indicates direction of shaft rotation. The circular arrow is to be viewed as above the connecting shaft or shafts. A circular arrow with a head at each extremity [Fig. 2.7(b)] indicates that the prime mover may be selectively driv-

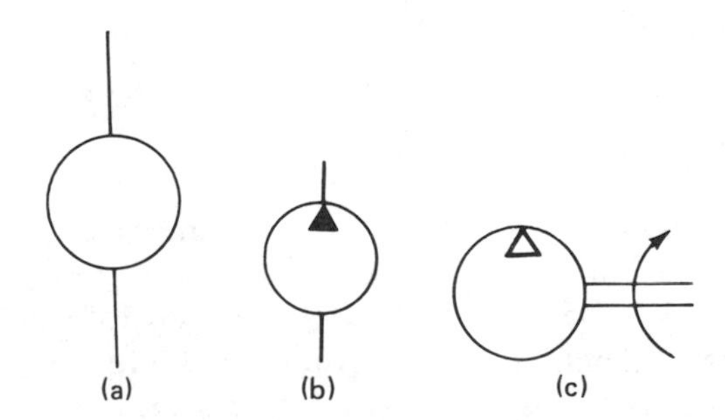

Figure 2.6 (a) Basic symbol for a pump or motor. (b) Solid equilateral triangle indicates liquid medium and unidirectional flow. (c) Hollow equilateral triangle indicates gas medium and direction of flow.

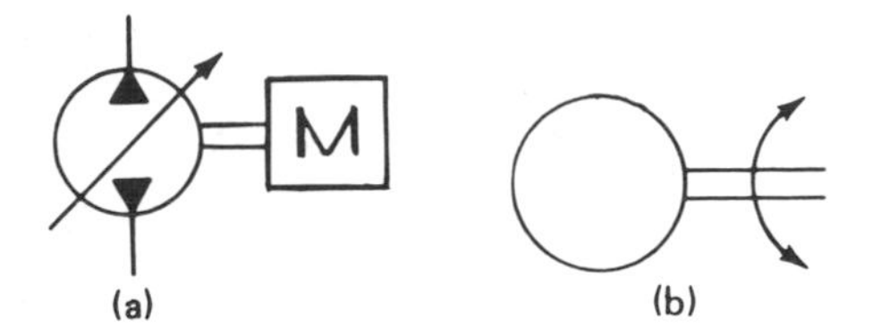

Figure 2.7 (a) Pump and prime mover connected. (b) Bidirectional drive.

en in either direction or that the direction of rotation is unimportant to the desired drive function.

Type of mechanical coupling of the prime mover output shaft and the pump input shaft is not indicated by the symbols. Likewise there is no indication of the pump/prime mover structure. A pump and prime mover sharing a common shaft is not usually differentiated by symbology.

2.3.3. Fixed Displacement Pump Symbols

A fixed displacement pump provides a predetermined flow at each increment of rotation of the drive shaft. Figure 2.8(a) indicates that flow passes through the pump as shown by the solid equilateral triangle. If the pump is designed to operate in two directions, the symbol incorporates two solid equilateral triangles [Fig. 2.8(b)]. Figure 2.8(c) shows two or more pumps sharing a common drive.

2.3.4. Flow Divider or Integrator

Two or more pumps sharing a common shaft may be employed to divide flow to a circuit in a predetermined pattern. The circuit may also be used to integrate flow from two or more sources into one common discharge.The pump could then function as either a motor or a pump according to the energy flow pattern. Each of the pump structures displace a predetermined flow per rev-

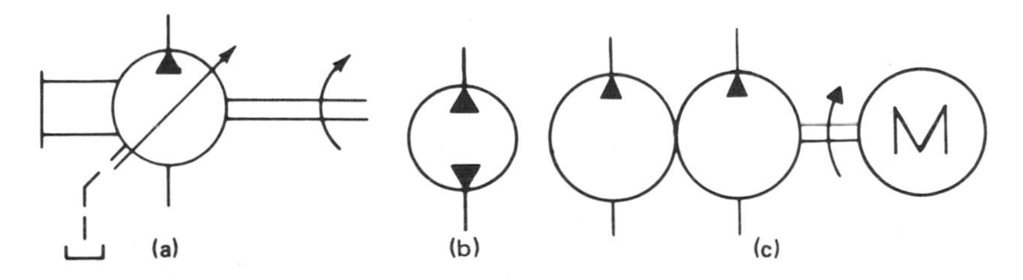

Figure 2.8 (a) Flow direction indicated by solid equilateral triangle. (b) Bidirectional flow. (c) Two or more pumps share a common drive.

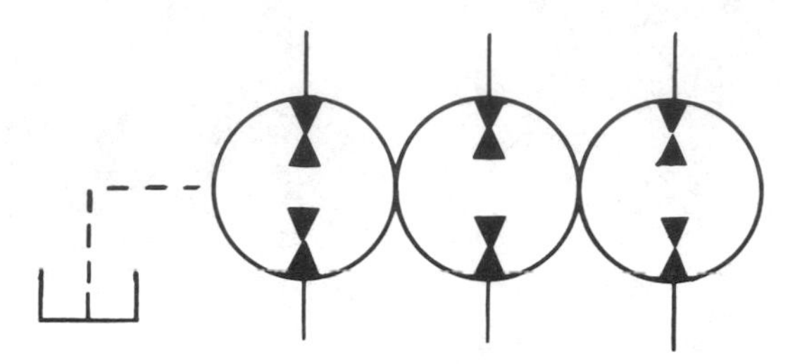

Figure 2.9 Flow divider.

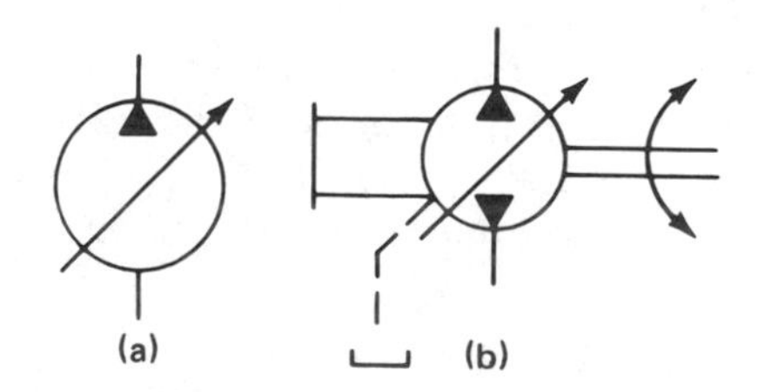

Figure 2.10 (a) Variable flow in one direction (simplified unidirectional). (b) Complete bidirectional.

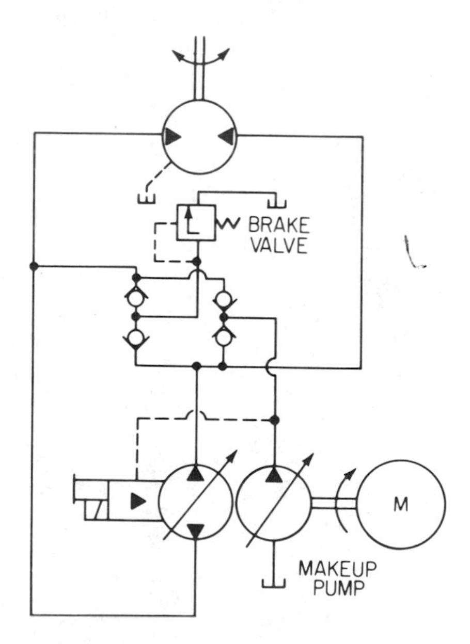

Figure 2.11 Auxiliary pump.

olution. Thus the energy need for each branch determines the function of the structure as a pump or hydraulic motor. Figure 2.9 illustrates the symbol as used for flow dividing function, flow integration, or reversible capabilities wherein a flow division is accomplished in one direction of rotation and a flow integration is accomplished in the opposite direction.

Displacement of the pumps used for a flow divider or integrator need not be equal. A predetermined relationship can be established. If pumps of different displacement are used for dividing or integrating flow, there will be proportional potential pressure and flow values related to the ratio of the displacement. The symbols will not show this potential intensification of pressure or amplification of volume delivered per revolution within the circuit.

2.3.5. Variable Displacement Pumps

A slash arrow added to the basic symbol for a fixed displacement pump indicates that the characteristics of the pump are now different and the displacement can be changed.

Figure 2.10(a) indicates potential variable flow in one direction. By adding the second equilateral triangle [Fig. 2.10(b)], a flow pattern in each direction is indicated and usually a neutral (no flow) condition is possible with the drive rotation in a predetermined pattern.

Addition of a fixed displacement pump to a common drive may require a uniform drive rotation of the prime mover. The addition of the fixed displacement pump (Fig. 2.11) provides potential lubrication flow, auxiliary supply source, control pressures, and/or liquid conditioning.

2.4. LINEAR AND ROTARY ACTUATORS

2.4.1. Hydraulic Motors

The equilateral triangle is pointed inward [Fig. 2.12(a)] to identify a rotary hydraulic fluid motor. A single equilateral triangle indicates unidirectional flow. Two equilateral triangles indicate bidirectional flow. Rotary hydraulic

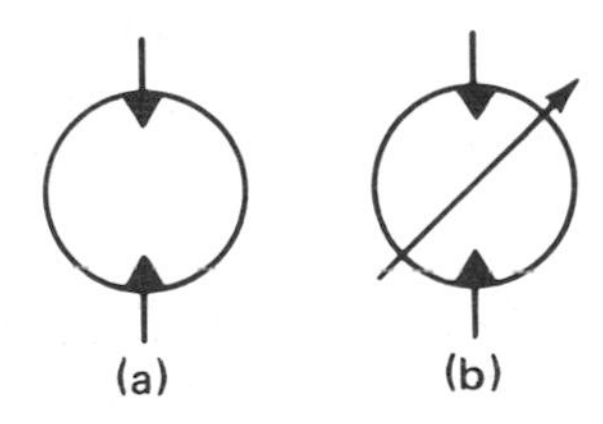

Figure 2.12 (a) Hydraulic motor. (b) Variable displacement hydraulic motor.

fluid motors can have the displacement varied. The slash line indicates this capability. If there is a desire to vary the motor displacement to provide a variable speed with a fixed or variable fluid supply, the symbol will include the slash arrow [Fig. 2.12(b)].

2.4.2. Limited Rotation Motors

Figure 2.13 identifies a rotary output hydraulic motor which has a limited rotational capability in each direction. The limited rotation motor may have characteristics similar to a linear hydraulic cylinder. The symbol does not indicate the design or construction.

2.4.3. Linear Actuators

Linear actuators are usually referred to as hydraulic cylinders or rams. Figure 2.14 illustrates the basic single action ram-type linear actuator. A ram of the

Figure 2.13 Motor with limited rotation.

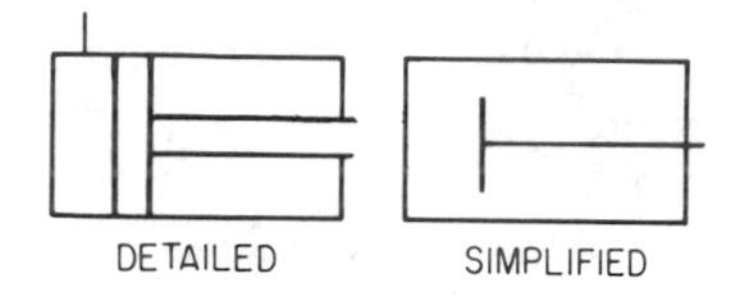

Figure 2.14 Single acting ram-type actuator.

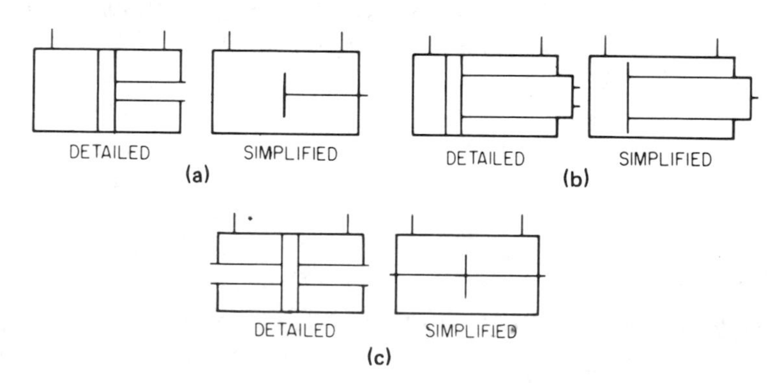

Figure 2.15 (a) Basic hydraulic double acting cylinder. (b) Oversize piston rod. (c) Double rod extension.

double-acting type may be retracted by pressurizing the cavity adjacent to the packing when a lip is provided on the ram assembly.

Basic double-acting cylinders with rod size determined by mechanical strength requirements are shown in Fig. 2.15(a). Modification of the basic cylinder to provide greater rod strength and/or less displacement at the rod end can be indicated by the symbol shown in Fig. 2.15(b). The intent is to show that there is a modification. Figure 2.15(c) shows a linear actuator with rod extending from both ends with equal displacement in each direction of travel.

2.5. PRESSURE CONTROL DEVICES

2.5.1. Relief or Safety Valves

A rectangular box with an arrow configuration (Fig. 2.16) represents a *normally-closed* two-way valve that is used for pressure level control and simultaneous control of flow direction. The arrow is below the input and output lines. The right angle tail does abut the input line indicating that input flow is contained within the valve but blocked from the output connection in the rest position. As the valve is actuated, the flow will be directed through the outlet connection in a predetermined pattern. A relief or safety valve senses pressure level upstream from the valve. Thus a pilot line is shown from the input line to the end of the rectangular box adjacent to the lower internal line illustrating flow (Fig. 2.17). Note that a zigzag line is placed opposite the pilot line indicating the presence of a mechanical spring which urges or biases the flow directing mechanism to the closed position when a pilot signal is not present.

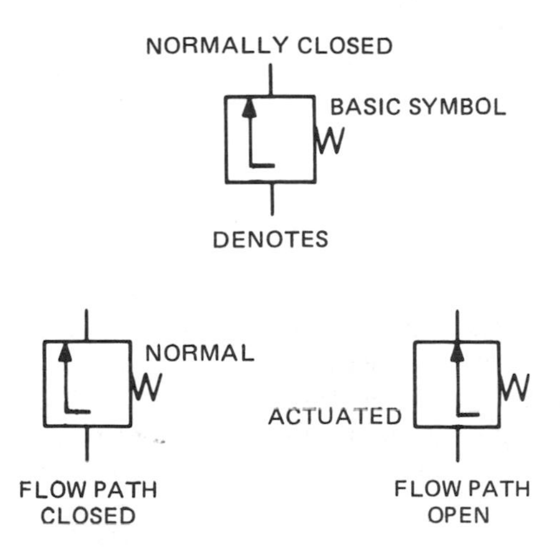

Figure 2.16 Normally-closed, two-way valve.

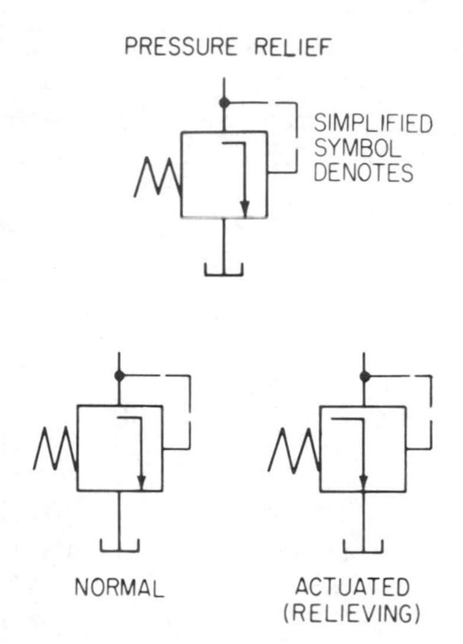

Figure 2.17 Relief valve.

A relief or safety valve is intended to direct flow to the reservoir at a predetermined pressure level. Thus it can be inferred that any drain function that is needed will be channeled internally to the outlet port of the valve with little resistance as it is returned to the tank or reservoir.

2.5.2. Sequence Valves

A sequence valve directs flow from the primary to the secondary port at a predetermined pressure level. The secondary port may be pressurized to a level approaching that of the primary value. Because of this an external drain connection to the reservoir is usually provided [Fig. 2.18(a)]. The input signal to the pilot area of the sequence valve can be internal from the upstream port [Fig. 2.18(a)] or it can be from an external source [Fig. 2.18(b)].

2.5.3. Counterbalance or Holding Valves

A sequence function is provided by a counterbalance or holding valve. In addition, the counterbalance or holding valve usually incorporates a return flow check valve. The pilot pressure source may be upstream or from an external point. Figure 2.18(c) shows the counterbalance or holding valve with external pilot and infers internal drain. Because a counterbalance or holding valve is always located in a cylinder line and contains a return flow check, there is no need for an external drain.

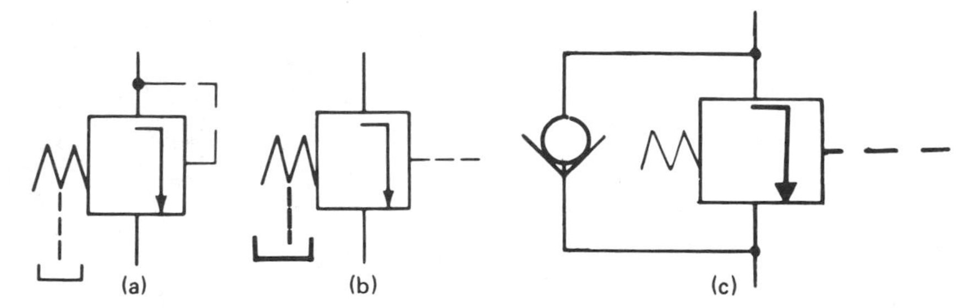

Figure 2.18 (a) Sequence valve—internal pilot, external drain. (b) Sequence valve—external pilot, external drain. (c) Holding valve—external pilot, internal drain.

2.5.4. Unloading Valves

An unloading valve is provided with an external pilot and internal drain (Fig. 2.19). An external drain might be used if the internal flows were very high and could adversely affect the control element through the internal drain connection.

2.5.5. Reducing Valves

A pressure level limit can be provided by a pressure reducing valve. The valve mechanism provides a normally-open passage from supply side to the output control area [Fig. 2.20(a)]. Operating signal is taken from the low pressure or control outlet of the valve. Drain is external. A check valve may be incorporated in the valve assembly for a free flow return [Fig. 2.20(b)].

2.5.6. Pressure Switches

Figure 2.21 illustrates an electric switch interface symbol. A pressure signal actuates the unit to make or break an electric contact. The electrical connec-

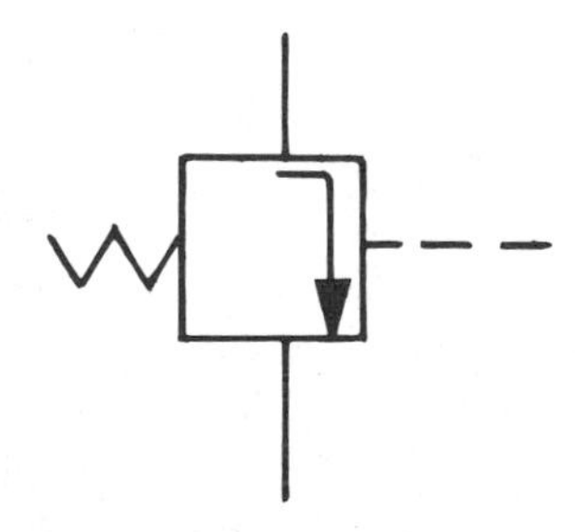

Figure 2.19 Unloading valve.

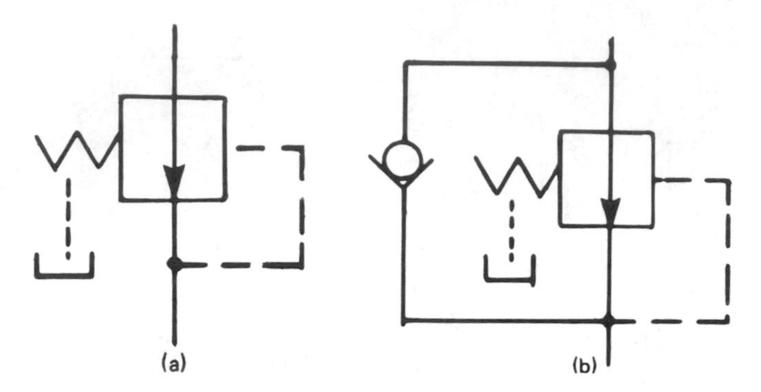

Figure 2.20 (a) Basic envelope—reducing valve. (b) Reducing valve with check.

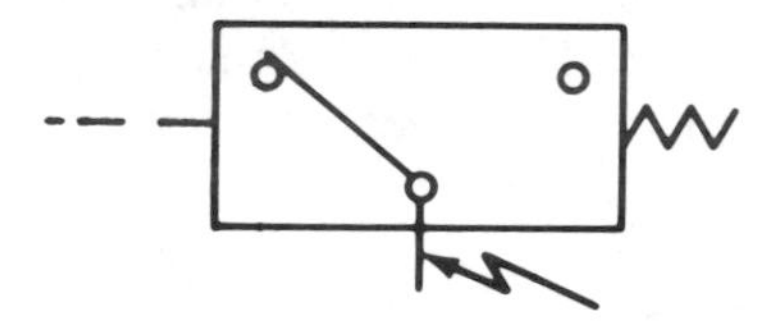

Figure 2.21 Pressure switch.

tion is shown when the unit is at the rest position without a signal applied at the pilot connection.

2.6. DIRECTIONAL CONTROL VALVES

Pressure level controls can also be considered as directional control valves. To understand the relationships we can look to the basics of the symbol structure.

2.6.1. Basic Symbols

The basic symbol for a hydraulic valve is a rectangle, termed the *valve envelope*. As with pumps and motors, the valve envelope represents the valve enclosure or body. Lines within the envelope show the flow directions between the valve inlet and outlet openings, termed *ports*. To show the change in flow conditions when the valve is opened or closed (*actuated*), two systems of symbols are used: (1) single envelope and (2) multiple element envelopes. Single-envelope symbols are used where only one flow path exists through the valve. Flow lines within the envelope indicate static conditions when the actuating signal is not applied. The arrow can be visualized as moving to show how the flow conditions and pressure are controlled as the valve is actuated.

Multiple-envelope symbols are used when more than one flow path exists through the valve. The envelopes can be visualized as being moved to show how flow paths change when the *valving element* in an envelope is shifted to its various positions.

Figure 2.22(a) shows a typical single envelope; multiple envelopes are shown in Fig. 2.22(b). Flow lines to the valve ports are added to the two types of basic valve envelopes in Fig. 2.22(c) and (d). Note in Fig. 2.22(d) that the ports are shown at the center envelope segment.

A variety of valve-port conditions are shown in Fig. 2.23. Thus, the double arrow in Fig. 2.23(a) denotes a locked or closed port in a single envelope where flow may be expected to pass in either direction. The blocked condition is indicated by the fact that the arrows do not line up with the lines coming to the single square. Figure 2.23(b) shows the same condition in a multiple-envelope valve. The single envelope signified that an infinite number of positions may be encountered between the closed (normal) and open positions, such as might be expected with a relief valve where the opening is a function of the orifice size that must be created to provide the needed pressure level in an automatic function.

The multiple envelope of Fig. 2.23(b) and (c) indicates a finite number of positions which the directing elements can be normally expected to assume. The normal position of Fig. 2.23(b) is indicated by the left block. The normal position of the unit, shown in Fig. 2.23(c), is the neutral position in which all the ports are blocked. This condition is referred to as a *closed center.* Shifted to conditions in the left block, the flow is in the direction indicated by the arrows. When shifted to the right, the flows are reversed. This is referred to as a *four-way function;* one input port can be directed to separate outlets or blocked, while a return line is directed to an opposite port in a predetermined pattern.

A through flow is indicated by the symbol of Fig. 2.23(d). An infinite number of positions of the flow-control elements from full-open to closed can be anticipated with this design. The two-element symbol of Fig. 2.23(e)

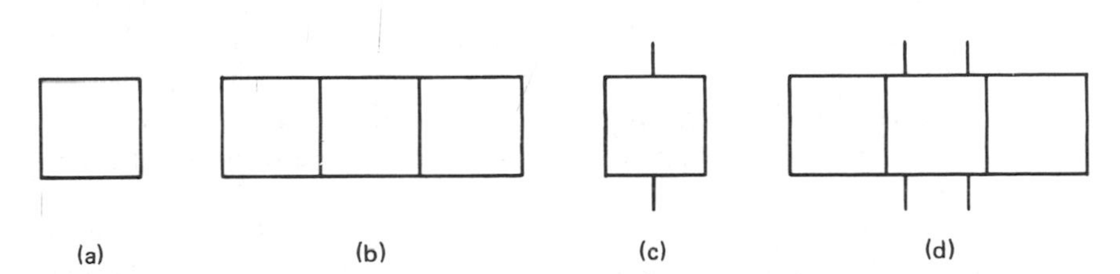

Figure 2.22 (a) Basic envelope. (b) Multiple basic envelope. (c) Ports attached to basic envelope. (d) Ports attached to normal multiple basic envelope.

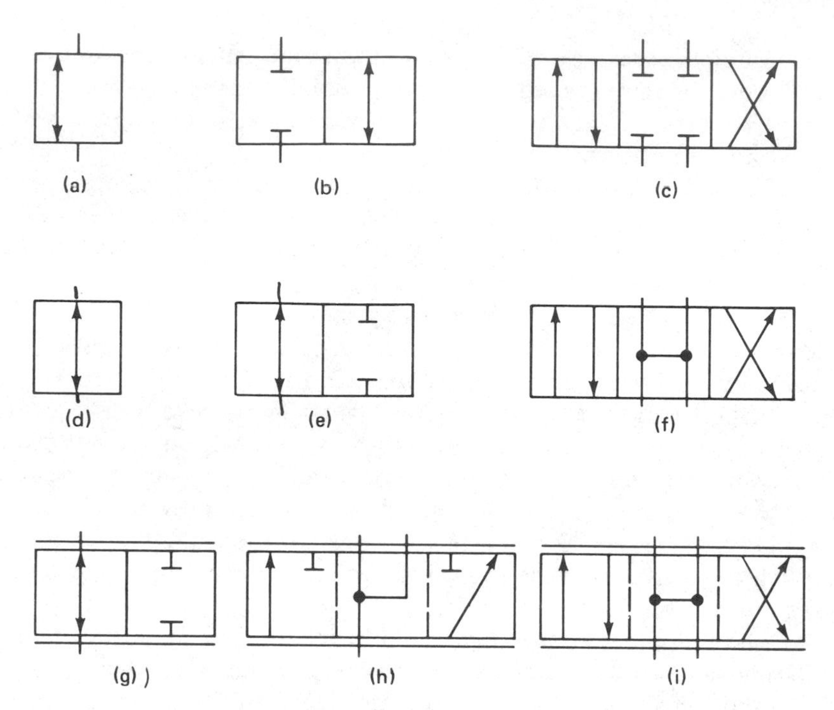

Figure 2.23 (a) Blocked port. (b) Blocked port, finite position. (c) Three-position, blocked ports in neutral, finite position. (d) Open port, infinite position. (e) Normally-open, finite position. (f) Open center, three finite positions. (g) Two-position, infinite crossover position. (h) Two-position, infinite crossover position, three-way diversion valve. (i) Two-position, infinite crossover, four-way valve with neutral fully open.

indicates two finite positions. The three-element symbol of Fig. 2.23 (f) indicates three finite positions, with the neutral position indicating an interconnection of all associated lines.

The two-element symbol of Fig. 2.23(g) is provided with parallel lines at the top and bottom. This signifies that it represents a valve with two finite positions; however, the nature of the application is such that it may be shifted to infinite positions in the process of moving from one position to the other. An example is a cam-operated two-way valve used to decelerate flow; the cam may follow a random pattern in the functional movement from open to closed position.

The diversion valve of Fig. 2.23(h) indicates that all ports are interconnected as the flow-directing element is moved. The dashed lines between the blocks indicate that the neutral position is not a finite one. The information contained in the center block indicates only the conditions as the valve elements are shifted from one finite position to the other. The open-center condition of the symbol shown in Fig. 2.23(i) would not normally be shown in a two-block symbol. Addition of the third position between the dashed lines provides the needed data without indicating three finite positions.

Figure 2.24(a) summarizes the basic symbol, normal, and actuated conditions for a normally-open, single-envelope valve. Figure 2.24(b) shows an alternate arrow design which indicates that one port is always connected to one flow path, whether normal or actuated.

2.6.2. Check Valves

The symbol for a *check valve* is shown in Fig. 2.25(a). Flow in this valve is from left to right. Do not confuse this symbol with an arrowhead. The circle represents the moving element of the check valve, often a *ball,* and the slant lines represent the *seat,* against which the ball is pressed when the valve shuts. Figure 2.25(b) shows the check valve with a bias spring. The valve opens if the inlet pressure is greater than the outlet pressure plus the spring pressure.

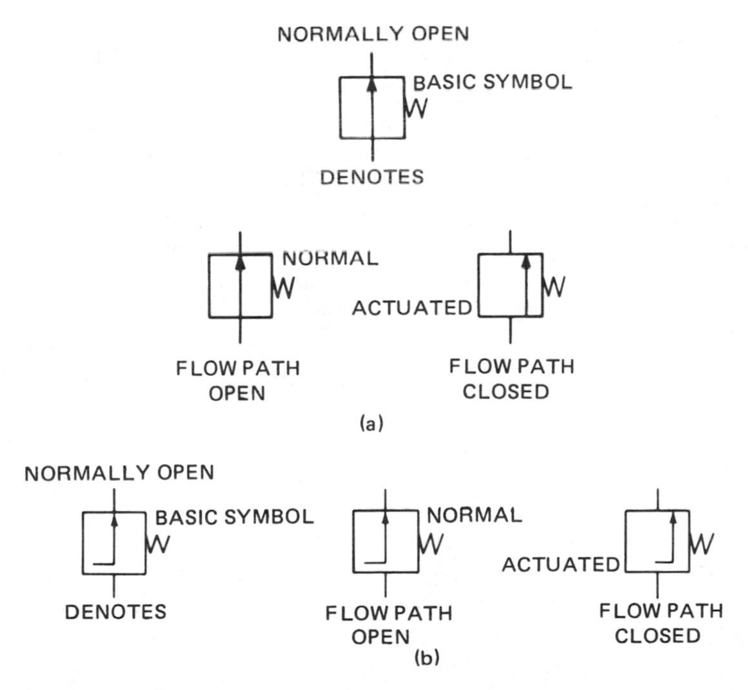

Figure 2.24 (a) Normally-open, single-envelope valve. (b) Alternate arrow design to indicate that one port is always connected, whether actuated or not.

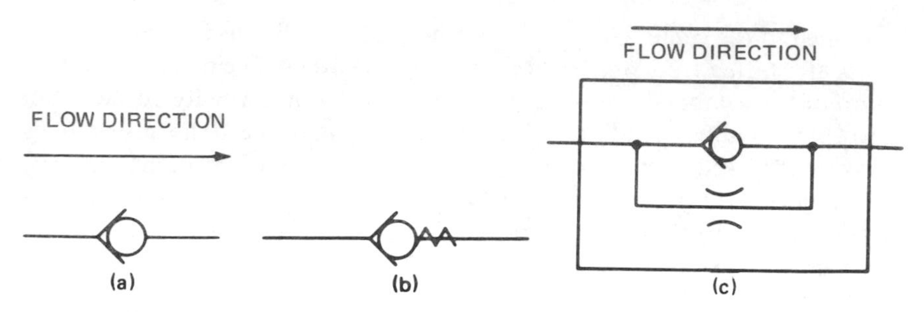

Figure 2.25 (a) Check valve—free, poppet opens if the inlet pressure is higher than the outlet pressure. (b) Spring loaded, opens if the inlet pressure is greater than the outlet pressure plus the bias spring value. (c) Unit allowing free flow in one direction but restricted flow in the other direction.

In Fig. 2.25(c) the check valve is fitted with an orifice. The box symbol provides an indication of the composite nature of the device and shows the fluid-energy path. The fluid flows freely from left to right through the check valve. If the check valve shuts and the fluid flow reverses itself, the orifice will restrict the amount of fluid passing from right to left.

Two *pilot-operated check-valve* symbols are shown in Fig. 2.26(a) and (b). This type of valve has an internal piston mechanism which may either force the valve open or hold it closed, depending on the design. The dashed line in Fig. 2.26(a) represents the pilot line used to conduct fluid to the piston mechanism to hold the check valve open. A drain line may be shown leading from the lower part of the enclosure. A baffle device is designed for certain models. The cavity between the baffle and upper surface of the operating piston must be drained. Generally, this drain is terminated above the fluid level in the reservoir. In Fig. 2.26(b) the pilot line is arranged to hold the valve closed. When pilot pressure is *relaxed* (relieved) in this valve, fluid may pass in either direction through the valve in certain designs. In other designs the check will perform in a conventional manner until pilot pressure is applied. At that time the check is held closed so that flow is not possible in either direction. Releasing the pilot pressure then permits the usual function.

2.6.3. Flow Regulating Valves

A manual shutoff valve of either the *gate* or *globe* type is represented by a symbol like that in Fig. 2.26(c). This type of valve is *not* intended for control of rate of flow in an infinite adjustment pattern. *Ball-type* manual shutoff valves are also widely used in similar functional activities as the gate and globe. For several types of *variable-flow-rate-control valves,* a symbol similar to that for an orifice is used. A slash arrow is drawn across the symbol [Fig.

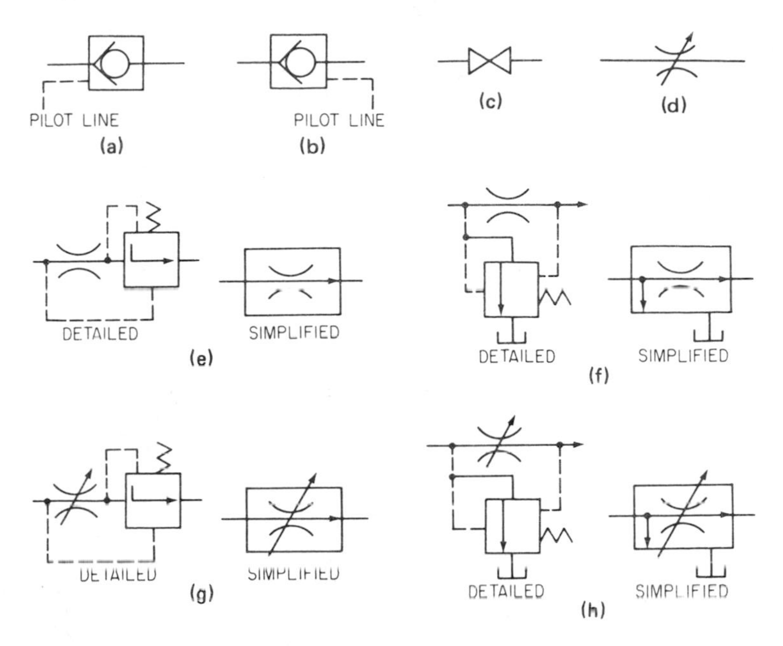

Figure 2.26 (a) Pilot check valve—pressure applied to open valve. (b) Pressure relaxed to open. (c) Manual shutoff. (d) Variable flow control valve. (e) Pressure compensated flow-control valve with fixed orifice. (f) With fixed output and relief port to reservoir. (g) With variable orifice. (h) With variable output and relief port to reservoir.

2.26(d)], indicating that flow can be varied by changing the orifice size. Pressure compensation is indicated by the symbols in Fig. 2.26(e)–(h). Variations in inlet pressure do not affect the rate of flow.

2.6.4. Multiple-envelope Valves

In this type of valve the basic symbol consists of (1) an envelope for each operating position of the valve, (2) internal flow paths for each valve position, (3) arrows indicating flow direction through the valve, and (4) external ports at the normal or neutral position.

Figure 2.27(a) shows a *two-position, three-connection valve* in its *neutral* or *rest* position. The external ports are always shown in the neutral or rest position in the valve symbol. Actuated, this valve is shown as in Fig. 2.27(b). Note that the port is shown as connected to the flow path represented by the arrowhead in the right-hand box. Two-position, four-connection valves vary only in that the fourth connection is added, as shown in Fig. 2.27(c) and (d). Three-position, four-connection valves have three working positions: normal [Fig.

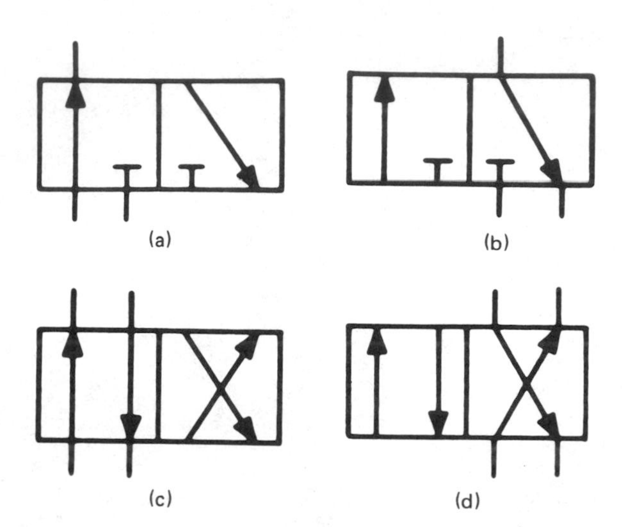

Figure 2.27 (a) Normal position of a two-position, three-connection, three-way valve. (b) Actuated position of a two-position, three-connection valve. (c) Normal position of a two-position, four-connection valve. (d) Actuated position of a two-position, four-connection valve.

2.28(a)]; actuated left [Fig. 2.28(b)]; and actuated right [Fig. 2.28(c)]. An additional two positions may be shown [Fig. 2.28(d)] with dashed dividers, indicating that the inner squares show a transitional condition. Thus, the symbol in Fig. 2.28(d) indicates that all valve ports are interconnected as the valve is shifted from neutral to an extreme position or from the completely shifted position back to neutral. Figure 2.28(a) displays this type of valve in the normal rest position where there is no need for showing the transitional conditions. In Fig. 2.28(b) the valve is in its actuated position with fluid flowing through the right-hand box. Flow is through the left-hand box in Fig. 2.28(c). Common neutral conditions for a variety of three-position, four-connection valves are shown in Fig. 2.29. Identifying letters for the various ports may vary with different manufacturers. The existing standards do not cover the identification of the ports, other than the U-shaped symbol indicating return to tank and the equilateral triangles indicating energy flow.

Pilot-operated, two-position, four-connection, spring-offset valves usually have some provision to drain the end cap containing the spring mechanism. Figure 2.30 shows the drain, with the pilot connection on the opposite end. External lines are connected to the box adjacent to the spring.

2.6.5. Types of Controls

When a spring is used in a control function, it is indicated by a wavy zigzag line [Fig. 2.31(a) and (b)]. It is placed so that one end touches the envelope.

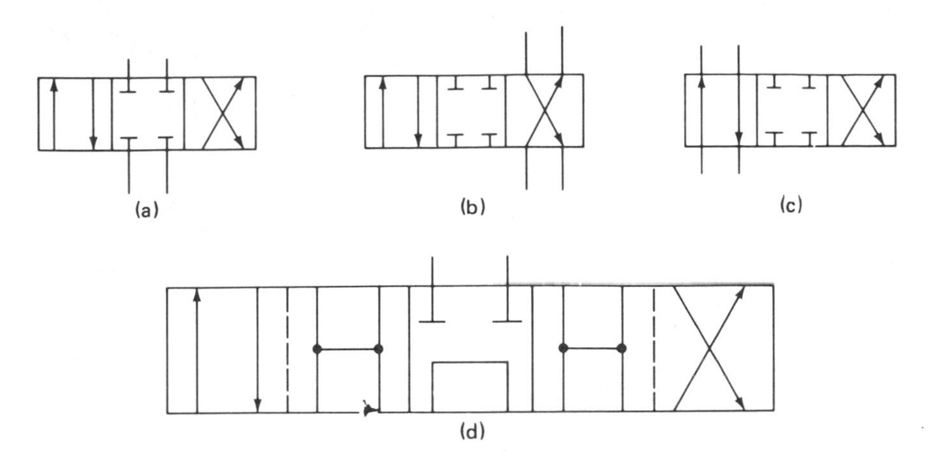

Figure 2.28 (a) Normal position of a three-position, four-connection, four-way directional control valve. (b) Actuated position of valve in *a*; envelope to the left. (c) Actuated position of valve in *a*; envelope to the right. (d) Dash dividing line to indicate transitory conditions as valve is shifted from neutral to extreme positions or from extreme positions back to neutral.

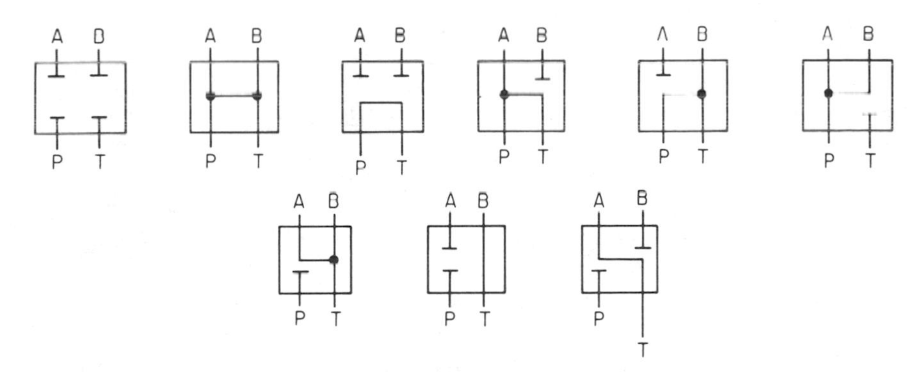

Figure 2.29 Four-way, directional control valve neutral configurations. Identifying letters vary with manufacturers.

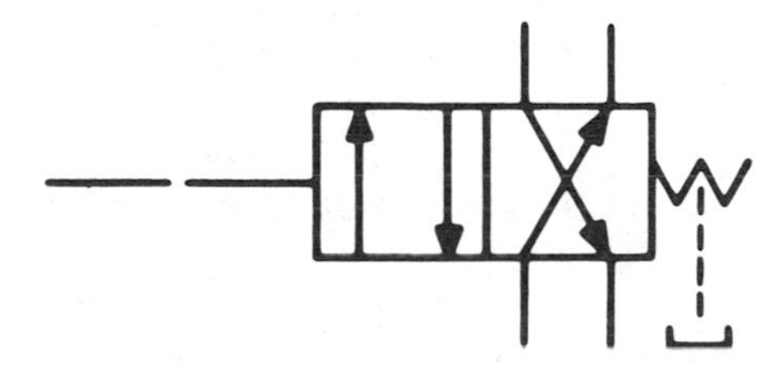

Figure 2.30 Two-position, four-connection, spring-offset, pilot-operated valve.

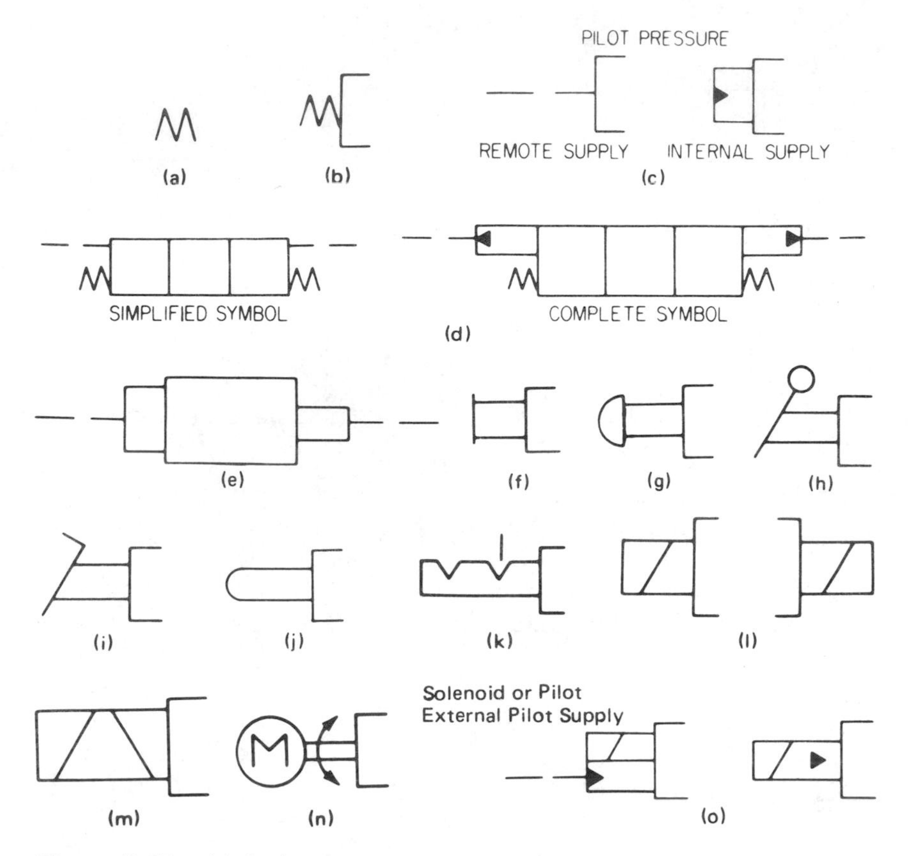

Figure 2.31 (a) Spring actuator. (b) Attached to box. (c) Basic pilot symbol. (d) Spring-centered pilot controlled assembly. (e) Pilot differential. In the symbol, the larger rectangle represents the larger area (i.e., the priority phase). (f) Manual operator used as general symbol without indication of specific type (i.e., foot, hand, leg, arm). (g) Push button. (h) Lever. (i) Pedal or treadle. (j) Plunger or tracer. (k) Detent (Show a notch for each detent in the actual component being symbolized. A short line indicates which detent is in use. Detent may be positioned on symbol for drafting convenience). (l) Solenoid. (m) Solenoid with two windings operating in opposite directions. (n) Reversing motor. (o) (1) Solenoid OR pilot, (2) solenoid AND pilot directional valve.

Figure 2.31(c) shows the hyphenated line coming from an external source. A pilot function can result from either an application of or a release of pilot pressure. Figure 2.31(d) shows the equilateral triangle pointing away from the major symbol, indicating that the operation is by release of pressure. Pilot actuation by differential pressure is shown in Fig. 2.31(e).

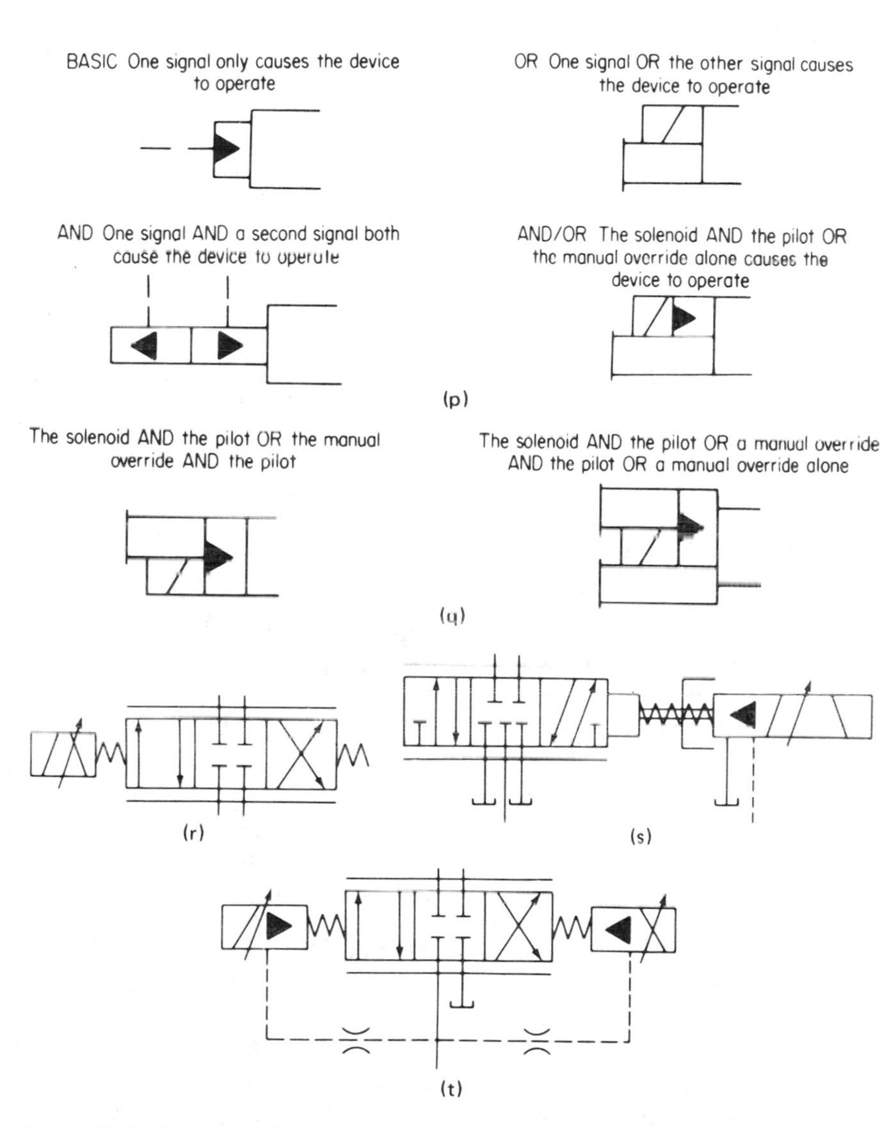

Figure 2.31 *(continued)* (p) Composite actuators – AND, OR, AND/OR. (q) Pilot, manual, and solenoid combination actuators. (r) Electrohydraulic servovalve: a unit which accepts an analog electrical signal and provides a similar analog fluid-power output. Single stage. (s) Two-stage with mechanical feedback and indirect pilot operation. (t) Two-stage with hydraulic feedback and indirect pilot operation.

The general symbol for manual operation is shown in Fig. 2.31(f), with the symbol for a push button in Fig. 2.31(g), lever in Fig. 2.31(h), and pedal or treadle in Fig. 2.31(i). Figure 2.31(j) indicates a plunger or tracer. A detent symbol [Fig. 2.31(k)] is most generally used with mechanically or pilot-operated valves. Some small-size pilot valves with dual-solenoid operation employ light detents to prevent spool shift because of machine vibration. The detent mechanically holds the flow-directing mechanism in the desired position until a specific force value is applied to cause movement from that position to another finite position. Individual solenoids [Fig. 2.31(l)] or a dual solenoid, with two windings operating in opposite directions [Fig. 2.31(m)] or reversing motor [Fig. 2.31(n)], provide electrical signal application. Combination of actuators are common. Applications may dictate use of solenoid OR pilot or solenoid AND pilot actuation [Fig. 2.31(o)]. AND, OR, AND/OR symbols are shown in Fig. 2.31(p) and (q). Electrohydraulic servovalves are shown in Fig. 2.31(r)–(t).

2.7. MISCELLANEOUS UNITS

Figure 2.32(a) shows the symbol for a gauge for pressure or vacuum. Figure 2.32(b) shows the symbol for a temperature-indicating gauge. The sensing bulb may be shown at a remote point if it assists in clarifying the circuit details. Flow meters [Fig. 2.32(c)] are of two types. *Rate* is indicated by the left symbol. The symbol at the right indicates a *totalizing meter* such as those used to indicate water consumption in homes.

Rotating joints in fluid conducting lines are indicated by the symbol in Fig. 2.32(d) when only one line is within the assembly. Two or more lines are shown as in Fig. 2.32(e), which also indicates a drain from the seal assembly.

The top accumulator symbol in Fig. 2.32(f) is used worldwide as a single symbol to designate any type of accumulator. American National Standards Institute graphical symbols also use the symbol of Fig. 2.32(f) as a general-purpose symbol. They, however, also recommend use of more explicit symbols such as those shown in Fig. 2.32(f) for spring-loaded, gas-charged, and weighted accumulators. The symbol for a spring-loaded accumulator encloses a zigzag and parting line to indicate use of a spring as an energy-storage medium. The symbol for a pressurized-gas accumulator (hollow equilateral triangle plus dividing line) usually indicates an inert gas as the energy-storage medium. A small rectangle on the dividing line is used to indicate a weighted-type accumulator. The symbol for a receiver in Fig. 2.32(f) is usually associated with compressed air. It may also be used to show an auxiliary tank connected to a gas-charged accumulator to provide a larger quantity of gas under pressure thereby increasing the energy-storage capacity.

Symbols for fluid conditioners employ a square resting on a corner with the associated major lines connected at the corners on a horizontal plane, as in Fig. 2.32(g).

The symbol for a heater [Fig. 2.32(h)] shows the introduction of energy into the major line. The cooler symbol [Fig. 2.32(i)] indicates heat dissipation or energy flowing from the major line. Dual arrows [Fig. 2.32(j)] indicate that energy can flow in either direction. A heater, coolor, or temperature-controller symbol can be provided with external lines, indicating whether transfer is by means of gas or liquid. A solid equilaterial triangle in a line, terminating at the midpoint on the diagonal of a square, indicates the medium is a liquid; a hollow equilateral triangle indicates a gas or compressed-air or a radiatory-type heat exchanger, such as that used on conventional, internal-combustion, water-cooled engines.

The symbol for a filter or strainer employs a dashed line bisecting the square [Fig. 2.32(k)]. No attempt is made to differentiate between the various types of filters or strainers. The position of the symbol in the circuit will usually provide a clue as to the type. Strainers are found most often on pump-suction lines, where introduction of major, large-size contaminants into the pump is prevented by the strainer. Filters may be found in any part of a circuit. However, machine tool circuits often employ high-pressure, in-line filters just ahead of critical components. Mobile equipment may use return-line filters most effectively because of the expected motion of the tank and resulting agitation of the fluid. There are no fixed rules as to strainer and filter usage, so the material list on a drawing may be needed to indicate the specific information.

Much pneumatic equipment is used in association with hydraulic systems. Because of this, it is usually easy to find pneumatic conditioning device symbols on predominantly hydraulic circuit drawings. Pneumatic directional-control valves, used, employ symbols similar to the equivalent hydraulic function. Hollow, equilateral triangles provide the clue as to the fluid medium. Figure 2.32(l) shows a separator used to remove moisture from pneumatic lines. The symbol of Fig. 2.32(m) illustrates a combination filter and separator for pneumatic lines. A desiccator, or chemical drier, is shown in Fig. 2.32(n).

Pneumatic lines usually require the introduction of a lubricant into the line to minimize rust and corrosion. The oil may also serve to assist in lubricating operating devices in the circuit, such as air-piloted hydraulic valves. Figure 2.32(o) shows the units with and without a manual drain. Figure 2.32(p) shows the symbol for an electric-drive motor. The symbol of Fig. 2.32(q) illustrates a heat engine, such as a steam, diesel, or gasoline engine.

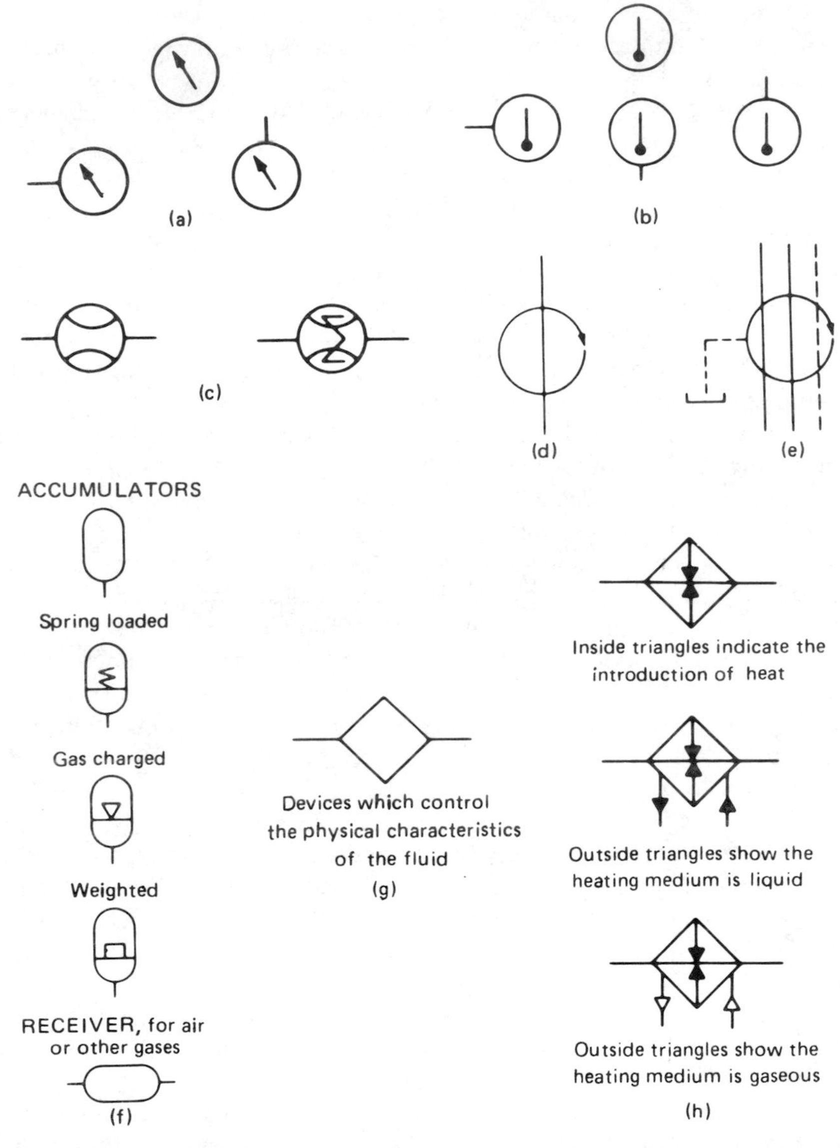

Figure 2.32 (a) Pressure or vacuum gauge indicating and recording pressure. (b) Temperature gauge. (c) Flow meters: left, flow rate; right, totalizing. (d) Rotating connection with one flow path. (e) Rotating connection with two major working lines, pilot line, and drain. (f) Accumulators and air receiver. (g) Fluid conditioner. (h) Heat exchange heater.

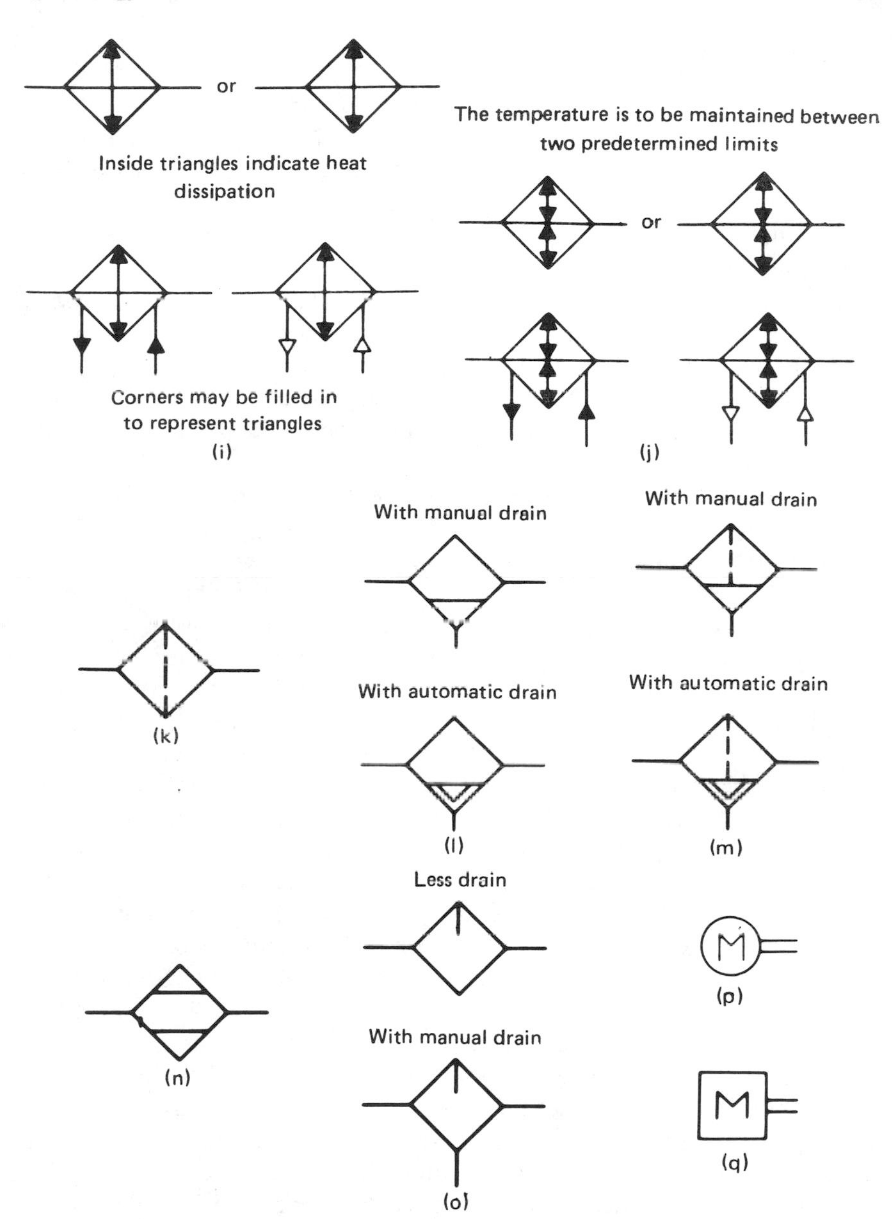

Figure 2.32 *(continued)* (i) Cooler. (j) Temperature controller. (k) Filter or strainer. (l) Separator. (m) Filter separator. (n) Desiccator (Chemical drier). (o) Air-line lubricator. (p) Electric motor. (q) Heat engine (Internal Combustion Engine).

2.8. COMPOSITE SYMBOLS

A *component enclosure* [Fig. 2.33(a)] may surround a composite symbol or a group of symbols to represent an *assembly*. It is used to convey more information about component connections and functions. The enclosure indicates the limits of the component or assembly. External ports are assumed to be on the enclosure line and indicate connections to the components. Flow lines cross the enclosure line without loops or dots.

In a typical component enclosure, one dash approximately 20 linewidths long is separated from the continuing line by a space of approximately five linewidths. The solid connecting line running lengthwise or around corners should be approximately 10 dash-lengths long. This can be varied but should include at least one dash to each side, top, and bottom. This enclosure is often referred to as a *centerline-type enclosure*.

An enclosure containing two pumps on a common shaft with a common suction line and independent discharge lines is shown in Fig. 2.33(b). An enclosure containing two motors of equal size with a common shaft used to equalize or split the flow of fluid can be shown in a similar manner [Fig. 2.33(c)]. The drain may be included in some designs if the motor case is not designed for pressures equal to the working pressure.

Composite symbols, such as those in Fig. 2.33(d), can be easily interpreted if a logical study pattern is followed. The best way to do this is to trace each line from its source to its endpoint. To start, follow the suction line from the tank to the pump. Starting from the pump opposite the rotating shaft (left pump), note that there is a working line passing through the enclosure to the circuit. Immediately beyond the pump discharge line is a line connected to a check valve. The position of the check valve indicates that the flow from this pump can never pass through the check valve. Thus, this point can be considered a dead end as far as the delivery of the pump is concerned. Directly above the pump at the left is the symbol for a relief valve. The discharge of this relief valve is piped through the enclosure to the tank. Below the relief valve is a pilot line going to a valve in the right-hand pump circuit. This pilot line usually provides only a signal, and movement of fluid in it is small. The pilot line will also be a dead end when the signal to this valve is completed. The left-hand pump, opposite the drive shaft, is provided with a circuit that can normally supply all machine pressure and flow requirements, but at certain times it may not provide sufficient flow for a special required speed. The extra pump is included in most circuits of this type to provide additional speed through added volume of fluid. The discharge of the right-hand pump is connected to the input of a normally-closed, two-way valve. Between the pump discharge and the inlet of the two-way valve, a connection is provided to the inlet side of a check valve. This check valve permits discharge of the right-hand pump to connect with the delivery of the left-hand pump, combining the volume of both pumps for delivery to the circuit. When pressure signal of the left-hand pump becomes high enough, it causes the pilot-operated, two-way valve to open, permitting the right-hand pump discharge to pass freely through this valve back to the tank. This return flow combines with the discharge from the relief valve and passes through the enclosure to the tank.

The symbol in Fig. 2.33(d) does not necessarily indicate precise values of the fluid pressure in the various parts of the circuit, but certain relationships can be assumed. The highest fluid pressure will be developed by the left-hand pump. The piloted two-way valve must be adjusted to open at a pressure *less* than the relief-valve opening pressure. If this adjustment is incorrect and the pressure required to open the piloted two-way valve exceeds that of the relief valve, then the delivery of both pumps will combine and pass through the relief valve to the tank when the circuit is not accepting fluid under pressure. The testing station shown provides a place to install a gauge so that the point at which the right hand pump starts to divert fluid to the reservoir can be checked and adjusted if necessary.

By tracing each line in this way and by knowing the basic symbols used, a complete understanding of all functional operations of a hydraulic circuit can be obtained. Figure 2.33(e) shows a flow-and-relief circuit enclosure. Trace its various parts in the same way as described above.

Figure 2.33(f) illustrates the composite and simplified symbols for filters, regulators, and lubricators in an assembly that serves to clean, lubricate, and regulate air diverted from a central system to be integrated into a fluid-*power* circuit.

Stackable valves, such as those used widely in mobile hydraulic circuits, may include many functional units in one manifolded assembly. Note the components included in the circuit of Fig. 2.33(g). The first two directional-control valves employed for major flow directing are in series. The last three are in a parallel arrangement. All pilot valves are in a parallel supply and return to tank circuit. Filtering is accomplished within the return to the tank line.

2.9. SUMMARY

Fluid power circuits using graphical symbols are international in nature. They provide clear-cut circuit information, regardless of language barriers. In a hydraulic circuit diagram, every part of a circuit must be clearly shown, with a minimum need for auxiliary notes. Symbols are available for most commercial components. By following basic rules, it is possible to create easily understood symbols for special applications.

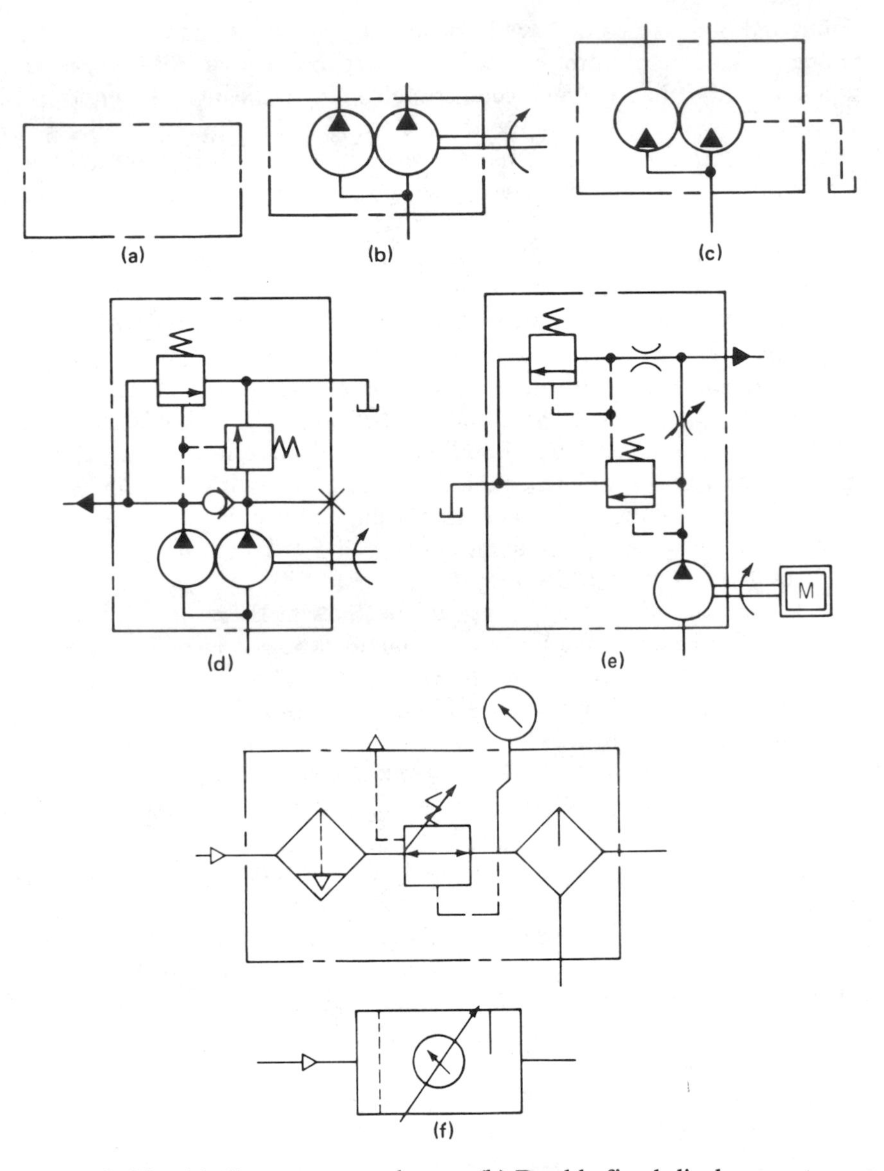

Figure 2.33 (a) Component enclosure. (b) Double fixed-displacement pump with one inlet and separate outlets. (c) Double fixed-displacement motor with one inlet and separate outlets (This assembly is used as a flow divider). (d) Double pump with integral check valve, unloading and relief valves. (e) Single pump with integral variable-flow-rate control and overload relief valve. (f) Control assembly for air supply to machine circuit. (Upper figure complete, lower figure simplified symbol).

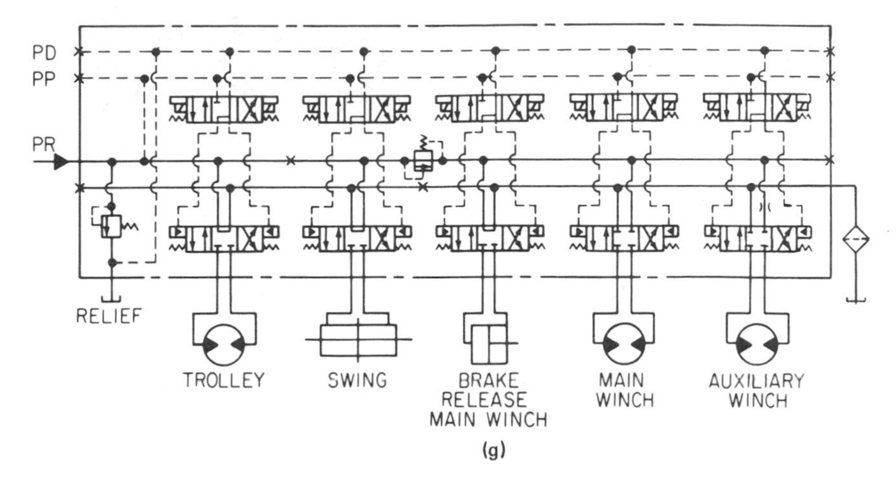

Figure 2.33 *(continued)* (g) Mobile stack valve assembly.

Symbols show connections, flow paths, and the function of the components represented.

Hydraulic symbols can indicate conditions occurring during transition from one flow-path arrangement to another. Symbols can also show if a device is capable of infinite positions or if it contains only finite working positions. A direct relationship is indicated between operator devices and the resulting flow paths. Symbols do not usually indicate construction or circuit values, such as pressure, flow rate, and temperature. Symbols do not indicate the location of ports, the direction of spool movement, or the position of control elements in actual components (Modified symbols may show some of these features in special units).

Hydraulic symbols may be rotated or reversed without altering their meaning except in the case of vented lines and lines to the reservoir, where the symbol must indicate whether the pipe terminates above or below the fluid level. Linewidth does not alter the meaning of symbols. A symbol may be drawn any suitable size. The symbol size may be varied for emphasis or clarity.

Where flow lines cross, a loop is used, except within a symbol envelope. The loop may be used in this case if clarity is improved. In multiple-envelope symbols, the flow condition, shown nearest a control symbol, takes place when that control is caused or permitted to actuate. Each symbol is drawn to show the normal or neutral condition of a component unless multiple circuit diagrams are furnished showing various phases of circuit operation. Arrows are used within symbol envelopes to show the direction of flow through the component in the application represented. Double-end arrows indicate reversing flow.

External ports are located where flow lines connect to the basic symbol, except where a component enclosure is used. External ports are at the intersection of flow lines and the component enclosure symbol, where the enclosure is used.

3 Pressure Control Valves

3.1. INTRODUCTION

Work is accomplished when any machine member is moved. The magnitude of the work is the criteria needed to design the mechanism to accomplish the desired movement and provide for the associated control force needs.

A hydraulic power transmission system uses pressure level control valves to provide several key control functions. To select valves to control these key functions we must know the following factors:

1. Available flow to perform work
2. Pressure levels to be provided
3. Rate of acceleration and deceleration
4. Desired maximum speed of actuation
5. Time frame for performing specific functions

The fifth item, time frame for performing specific functions, may be the key to choice of actuator size, pump capabilities, and then, the choice of pressure levels to be established.

Pressure level controls will fall into families associated with the function to be performed. We can divide these categories into six basic families and several subfamilies.

1. *Relief and/or Safety Valves*, first, limit the maximum system pressure which, in turn, protects the system components, piping, and tubing; and second, *limit the maximum output force* of the hydraulic system.

2. *Sequence Valves* are used to assure that one operation has been completed before another function is performed. They operate by *isolating the*

secondary circuit from the primary circuit until the set pressure is achieved. The flow of fluid is then sequenced from the first to the second circuit. Primary pressure must be maintained for the secondary function to be performed. Sequence valves *establish an order for the interaction of forces*.

3. *Counterbalance, overcenter, holding, or brake valves* are a broad range of pressure valves which control a load induced pressure to hold and control the motion of a load. This group of valves *provides balancing forces which prevent the load from running away* because of its own weight or because of inertia.

4. *Unloading valves* are usually used in circuits with two or more pumps or in circuits incorporating accumulators. The valve operates by sensing pressure in the system downstream of a check valve. Once a certain pressure level is obtained, the unloading valve unloads its pump to tank. The check valve isolates the unloaded pump from the rest of the system. System pressure is then maintained by the accumulator(s) or by a smaller volume high pressure pump. Unloading valves may be used to divert flow from the rod and of a cylinder at a predetermined pressure setting in a regenerative type circuit. It may be considered as a two-way valve when used in this mode.

5. *Reducing valves* are used to limit a certain branch of the hydraulic circuit to a pressure *lower* than the relief valve setting for the rest of the system. By reducing pressure in the secondary circuit, we can *independently limit the output force to that in the primary circuit*.

6. *Miscellaneous pressure sensing devices and combination valves* Signal sources can be sensed and measured by pressure switches, gauge mechanisms, and pressure transducers in a pattern that can affect the pressure levels in a hydraulic circuit. Combinations of valves in a common housing can minimize plumbing and reduce leakage potential.

A thorough knowledge of each of these families is essential to the task of developing a hydraulic circuit and choosing the best valve of the correct size and rating for a specific machine. Rating information may be available only from the component manufacturer through his published catalog data or by consulting with his sales personnel.

3.2. TYPES OF VALVES

3.2.1. Safety Valves

Safety valves, as the name implies, are used to limit the maximum pressure level within a portion of a hydraulic circuit to a level which will not cause damage to the circuit components or associated machinery. A pressure level can also be established and maintained by limiting flow and pressure to the controlled area within the circuit.

Common usage of the term safety valve designates a normally-closed, two-way valve (Fig. 3.1) which is designed to remain closed until an upstream

Figure 3.1 Safety valve: to open the valve, pressure acts against single stage poppet to overcome spring force.

pressure level reaches the value of the biasing mechanism urging the poppet against the seat.

Biasing Mechanisms

A biasing mechanism is the device used to hold the restrictive device against a mating seat to provide the fluid barrier. It can be a spring, external pressure source (hydraulic or pneumatic), or a weight. The area of the face of the movable restrictive member that is in contact with the pressurized fluid determines the force needed to maintain the closed condition prior to the relieving action.

Physical Weight and Gravity The force will be relatively constant if a weight is used to hold the movable member against the seat. A weighted device is dependent upon physical position. This limitation is unacceptable in most installations so a mechanical spring or a hydraulic or pneumatic force provides an acceptable alternative to the very dependable physical weight and gravity.

Direct Spring A spring is rated in pounds-per-inch (ppi) of deflection. Thus a spring must be installed with an adequate preload to provide the desired initial pressure level. The rate of deflection of a spring is dependent upon the type of wire used, the size of the wire, the number of coils, and the mean diameter.

To provide high pressure control a relatively large single spring may be needed. To reduce the physical space required two alternate designs are used.

Dual Springs If the design permits, additional springs may be nested together with alternately wound coils (right and left hand) so that they will not physically interlock. The springs can be designed to work in a predetermined parallel pattern.

Hydraulic Pilot Assist A second design uses pressurized liquid or gas to assist a basic bias spring [Fig. 3.2(a)]. The basic bias spring urges the movable member against the seat. The pressure needed to cause movement with the basic bias spring may be quite low. The cavity within which the basic spring is located, or a parallel piston assembly, may be employed to add hydraulic or pneumatic force to the basic spring.

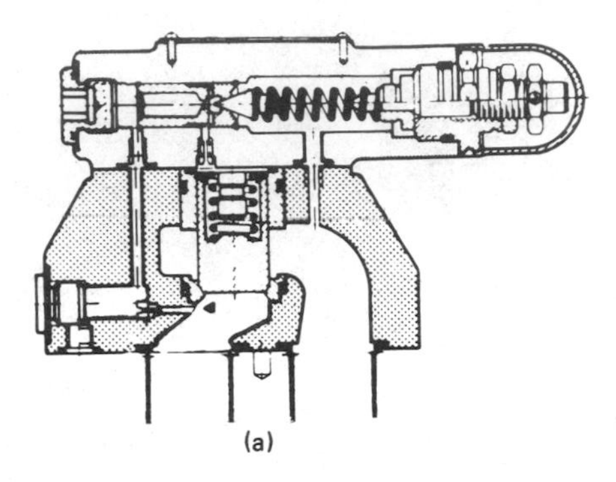

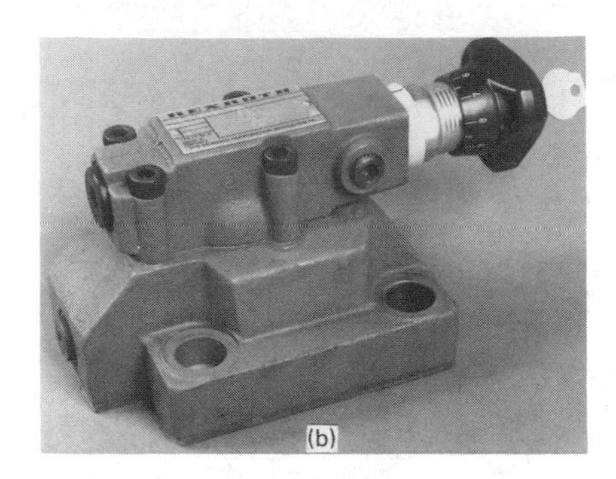

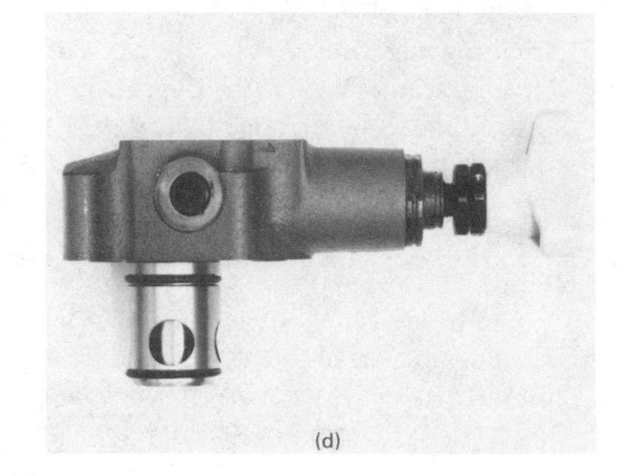

Figure 3.2 (a) Piloted relief valve. (b) Lockable relief valve. (c) Cartridge structure in threaded body. (d) Cartridge element and pilot cap. (Courtesy of The Rexroth Corporation, Bethlehem, Pennsylvania.)

The source of hydraulic pressure can be internal. A small hole connecting the chamber with the primary pressurized fluid source provides a supply of fluid which in turn reflects the increasing or decreasing pressures in the basic circuit until the levels reach the pilot control level at which time the maximum pressure is established. In this function a small auxiliary spring holds a poppet-type closure against a mating seat of small diameter.

The major closure cannot open against the primary bias spring and contained fluid until the pressure reaches the value needed to open the small auxiliary spring and poppet assembly. When fluid escapes through the auxiliary

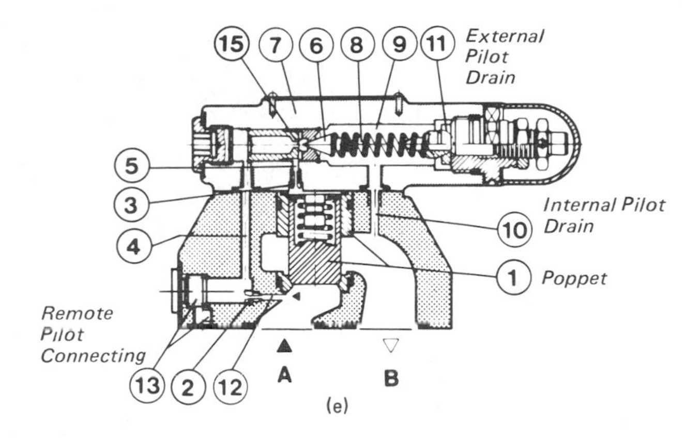

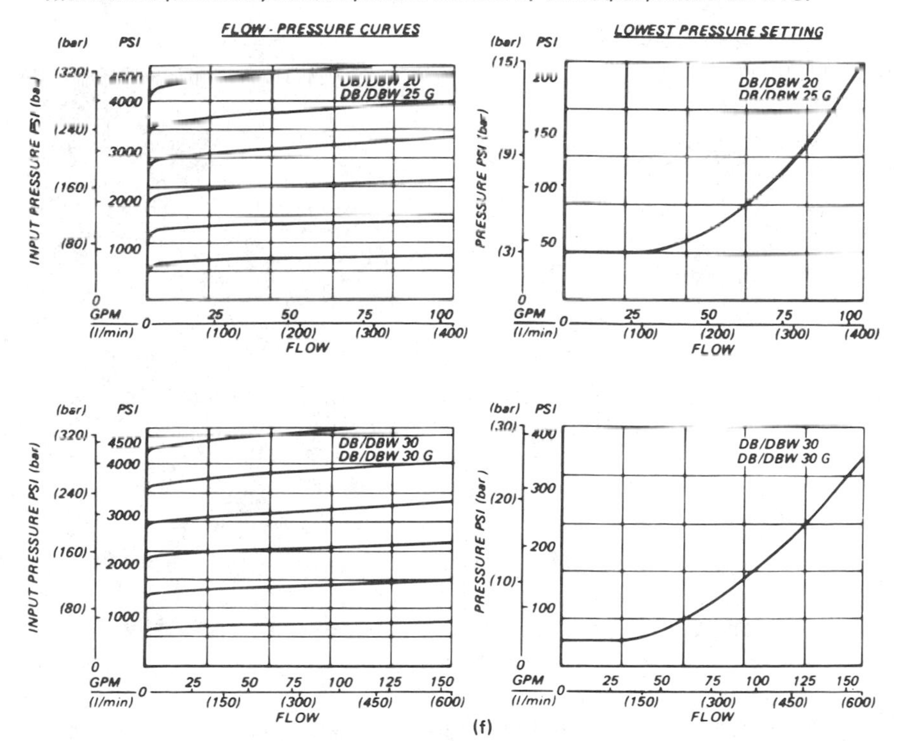

Figure 3.2 *(continued)* (e) Cutaway of pilot relief valve. (f) Typical performance curves for a piloted relief valve. (Courtesy of The Rexroth Corporation, Bethlehem, Pennsylvania.)

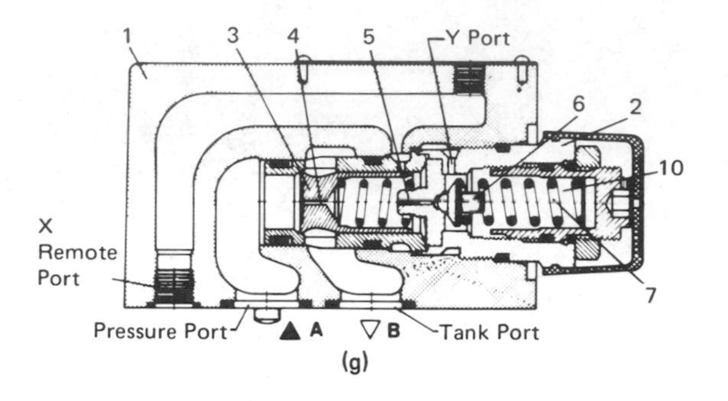

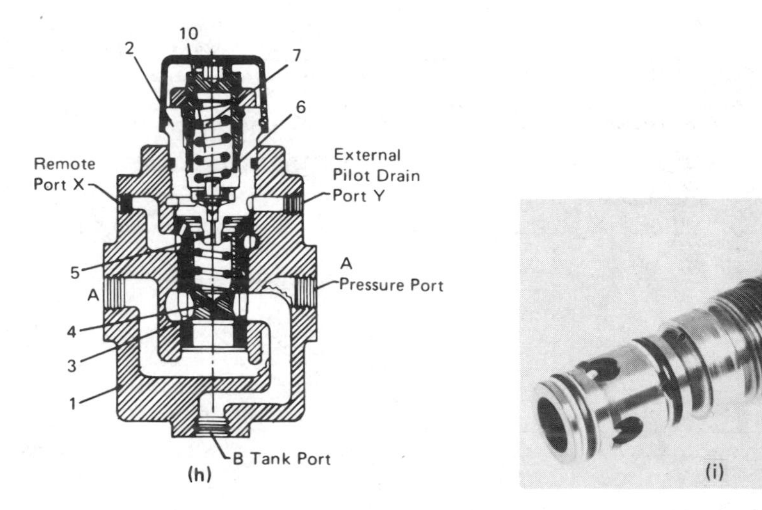

Figure 3.2 *(continued)* (g) Subplate mounted cartridge-type valve. (h) In-line threaded connection cartridge-type valve. (i) External view of cartridge structure. (Courtesy of The Rexroth Corporation, Bethlehem, Pennsylvania.)

poppet assembly at a rate greater than the pilot supply flow, the major closure can move in a servo-like fashion, providing a passage for the major flow. At this point a maximum pressure level is established. At this pressure level the valve diverts fluid to a secondary area or to the reservoir. Sufficient fluid is passed to an outlet area to prevent further build up of pressure within the major area.

The pilot-controlled main relief spool is not limited to any size relationship with the associated pilot valve or valves. Main relief valve assemblies can be in cartridge form so that they can be subplate mounted as in Fig. 3.2(b), provided with a threaded body as in Fig. 3.2(c), or furnished as a pilot and poppet assembly as shown in Fig. 3.2(d) for integration into other envelope structures. Figure 3.2(e) shows a cross section view of the complete relief assembly as housed for a subplate unit. The poppet assembly (1) is made up of a sleeve/seat structure, movable poppet, bias spring, and a hat-shaped metal insert in the center of the spring which reduces the oil volume above the main poppet in order to lower the capacitance and increase the valves' natural frequency. This makes the valve very quick to respond and stable. Orifice (2) accepts flow through channel (12) from the input port A. Remote pilot connection (13) can be connected with other pilot valves (can be at a remote location) to provide pressure levels less than that established by the master pilot. Pilot fluid passing through passage (4) is effective at the face of poppet six which is seated in sleeve/seat (15) and also effective on the upper face of poppet (1) passing through hole (5) and orifice (3). Adjustment (11) is provided with a mechanical stop to prevent overpressurization of spring (8). Spring (8) urges cone (6) against the seat. When pressure rises and cone (6) is forced from the seat, the resulting flow passes into chamber (9) and through passage (10) to B or tank. When the flow by cone (6) is greater than the flow through orifice (2), the pilot pressure at the top of piston (1) decreases until an unbalance occurs at which time a major flow is established from A to B.

The control flow across poppet (6) establishes the major flow value through the main poppet which can be of any practical size. Main poppet assemblies are catalogued in sizes to provide flows approaching 1000 gpm and pressures above 5000 psi.

The charts of Fig. 3.2(f) show the relative pressure/flow relationship and minimum pressures created by the main poppet bias spring and inherent resistance to flow.

Figure 3.2(g) and (h) shows how a similar structure can be integrated with the pilot/poppet asembly of Fig. 3.2(i) into a subplate unit [Fig. 3.2(g)] or a threaded connection assembly [Fig. 3.2(h)].

When the valve is used for safety or relief purposes, the secondary area is usually a pipe connected to the reservoir with minimum resistance to flow.

Function of a safety valve may be limited to protection in case of a machine malfunction. In such installations a simple device such as a ball urged against a seat by a compression spring may be quite adequate.

Sound Level

Sound level of a simple safety valve may be of little consequence. Sound level of a safety valve may be designed to be high to alert the attendant to the existing malfunction of the machine. A relief valve to establish a working pressure level is usually quiet in operation.

Adjustment

Safety valves may not be provided with pressure level adjustment devices (Fig. 3.1). If there is an adjustment device, it may be secured or hidden to avoid unauthorized adjustment. A wire and seal mechanism may be used to secure the cap covering the adjusting device so that tampering can be detected. Sometimes a keyed lock is provided at the pilot adjustment as shown in Fig. 3.2(b).

The usual safety valve will reset after the overload condition is remedied. Some adjustments provide a mechanical stop to prevent adjustment beyond a specific safe value, or a catalogued maximum pressure value.

Flow Fuses

Flow fuses are designed to protect a hydraulic system from loss of fluid if a line breaks. The valve stays open and allows flow between the system and an actuator at normal flows, but closes and shuts off all flow instantly if a line breaks and flow from the actuator exceeds the setting of the valve. This will lock the actuator in place and protect personnel in the area of the break.

A typical valve consists of a two-way structure that is biased by a mechanical spring to hold the valve open to pass a predetermined maximum flow.

A pressure signal at the inlet from the pump system passes through an orifice to the spring cavity at one end of the two-way valve spool. The cavity at the opposite end of the two-way valve spool is connected without restriction to the protected actuator port.

Should a line break between the pump and the flow fuse inlet, the pressure signal assisting the bias spring is negated and pressure created by the actuator will pilot the two-way valve spool to the closed position holding the actuator in a static position until the break is repaired and pressure is again available to assist the bias spring and permit the spool to move to the restricted open flow position.

3.2.2. Pressure Fuse

Frangible Disc

A frangible disc securely held in a housing connected to the pressurized line serves the function of a safety device (Fig. 3.3). The secondary line downstream from the disc is usually directed to the reservoir. At a predetermined pressure value the disc will rupture creating a passage from the pressurized line to the reservoir. It will be impossible to repressurize the line until the disc is replaced.

Kickdown with Automatic Reset

"Kickdown" relief and sequence (relief with external pilot drain) cartridges act like circuit breakers in an electrical circuit. They can unload a hydraulic system completely when the cartridge setting is reached, rather than holding and maintaining a fixed pressure. This protects machinery which could be damaged by overheating or overloading if held under pressure for any period of time.

Kickdown cartridges are specialized forms of pressure control valves. In use, they remain closed at any pressure below their setting. When the pressure

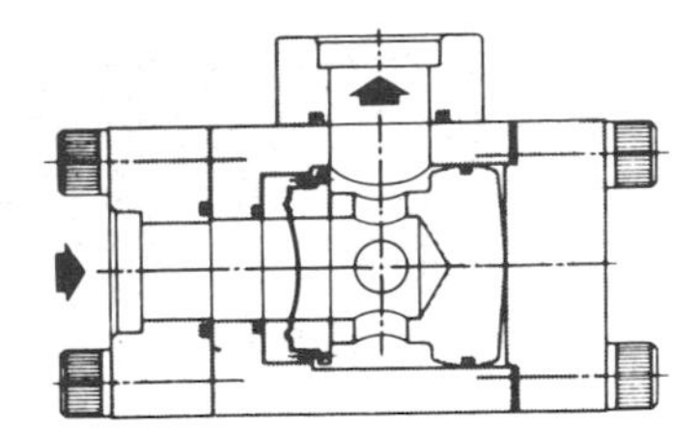

Figure 3.3 Hydraulic Fuse. (Courtesy of Double A Products Co., Manchester, Michigan.)

setting of the valve is reached, it opens completely and allows full flow through the valve at a very low-pressure drop.

This type of valve protects circuits where it is not desirable to maintain pressure on an actuator once the maximum pressure is reached. In drilling applications, for example, the machine designer may not want to allow the operator to continue to maintain full torque on a drill if the hole fills with chips. In this situation the valve will open at its pressure setting (maximum torque) and drop the system pressure to a predetermined low level. The operator would be required to reverse the drill, clearing the chips and, at the same time, allow the valve to reset for maximum torque on the next pass.

The kickdown sequence cartridge remains closed until the pressure in the primary circuit rises to the setting of the valve. At the set pressure, the valve opens, and will remain open even if conditions in the secondary circuit causes the system pressure to fall below the valve setting.

3.2.3. Combination Relief and Safety Valve

A relief and safety valve combination can be incorporated in a directional control valve so that the maximum pressure level will be established at the directional control valve pressure supply inlet (Fig. 3.4). This can be for reasons of economy of manufacture or because of the desire to localize the pressure level at the specific directional control valve and to simplify plumbing.

Relief and safety valves can be in cartridge form [Fig. 3.2(g–i) Fig. 3.5(a)], gasket mounted [Fig. 3.2(b), (e), (g)], or mounted in pipe lines with threaded connections [Fig. 3.2(c), Fig. 3.6]. Figure 3.5(b) shows the symbol for the cartridge relief valve whether in basic cartridge form or installed in an envelope.

Cross-Port Relief and/or Safety Valves

Relief and/or safety valves are frequently installed in the cylinder lines beyond a four-way directional control valve and ahead of the actuator [Fig. 3.7(a)]. Figure 3.7(b) shows the symbol.

Figure 3.4 Directional control with integral relief valve. (Courtesy of General Signal-Hydreco, Kalamazoo, Michigan.)

The secondary port from each of the relief valves is connected to the opposite cylinder line. Thus any localized pressure in the cylinder line will be limited and pressurized fluid will be directed to the lower pressure in the opposite cylinder line.

Cross-Port Relief With Anticavitation Check Valves

Actuators may have external drains so that fluid can be lost in this isolated portion of the circuit. A check valve network as shown in Fig. 3.8(a) and (b) will permit entry of make-up fluid.

Operating Pressure Levels

The pressure level at which a safety valve functions may be significantly different than the pressure at which it recloses. As an example—the release pressure may be 25% higher than the pressure at which it recloses. Thus the working pressure may be considerably less than the relief setting so that in the event of a malfunction the pressure will release at a safe value. As the

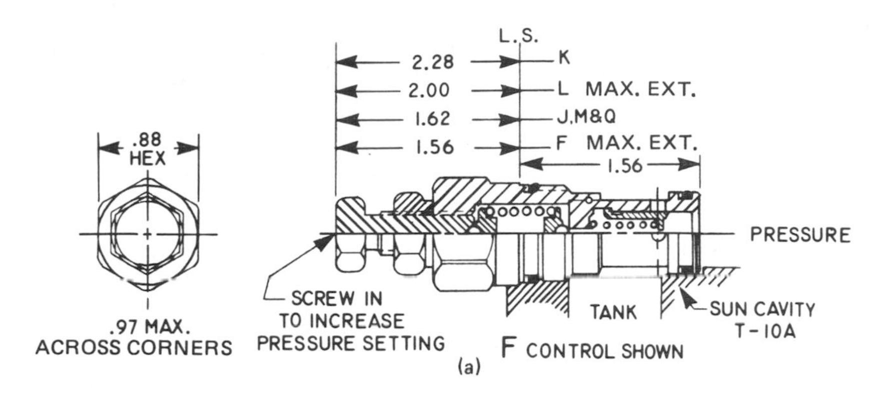

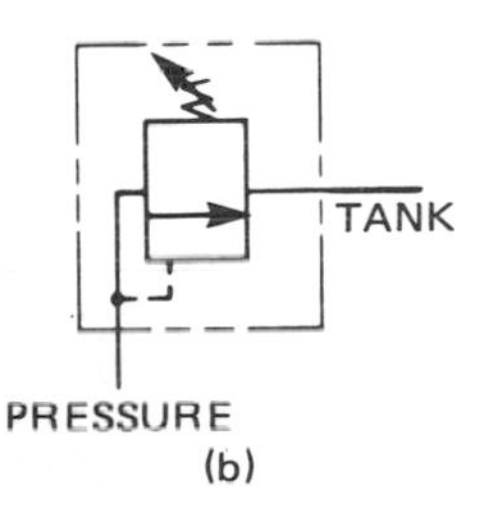

Figure 3.5 (a) Cartridge relief valve. (b) Symbol for cartridge relief valve. (Courtesy of Sun Hydraulics Corporation, Sarasota, Florida.)

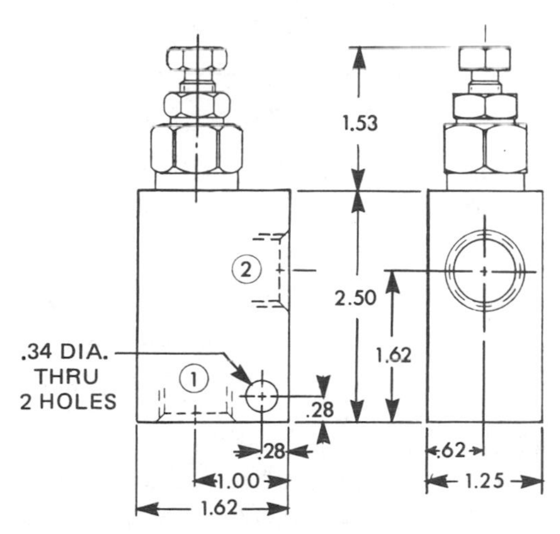

Figure 3.6 Relief valve cartridge installed in line-mounted body with threaded ports. (Courtesy of Sun Hydraulics Corporation, Sarasota, Florida.)

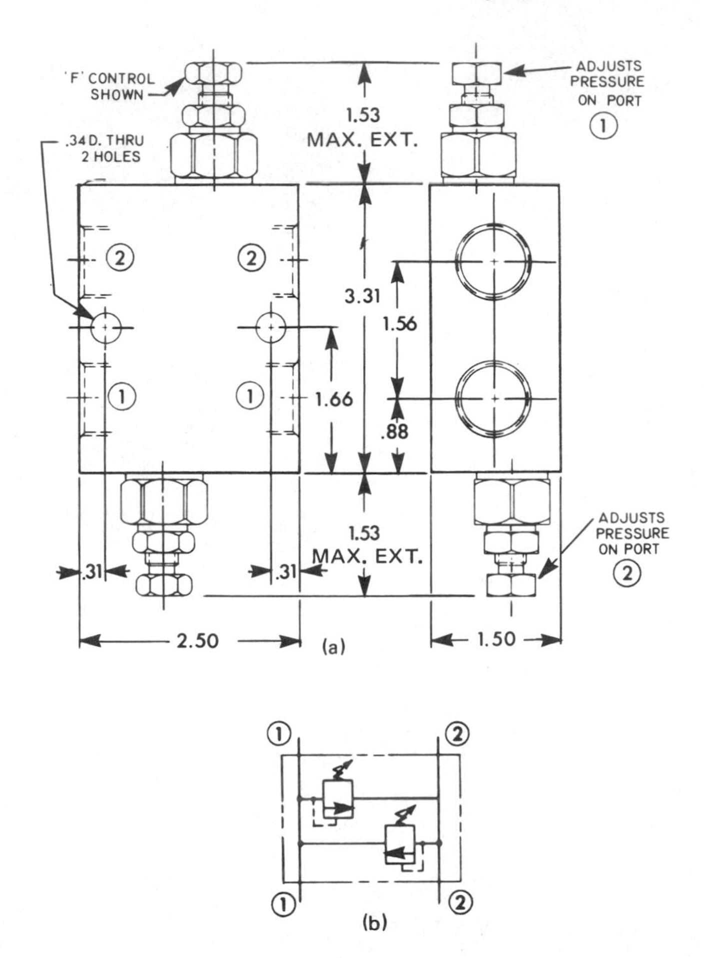

Figure 3.7 (a) Dual cross-port relief valve (b) Symbol for cross-port relief. (Courtesy of Sun Hydraulics Corporation, Sarasota, Florida.)

malfunction is corrected, the reset pressure is still greater than the normal working pressure so that working fluid is not lost through the safety valve.

A *relief valve* can be designed for a very narrow difference between the initial opening pressure and full open pressure value. The reset value may be less than 10% lower than the release pressure. Those relief valves designed with a close range between release and reset are usually of the piloted type which use a basic bias spring supplemented with a pilot valve and hydraulic pressure additive.

3.2.4. Piloted Relief Valves

The pilot valve need not be physically adjacent to the piloted main relief valve segment as mentioned earlier in this chapter. The pilot valve can be many feet away so long as it is connected with suitable pipe or tube. The resistance to flow in the pilot pipe is additive to the resistance created by the spring closure assembly in the piloted circuit.

Pilot Valves in Parallel

Any number of pilot valves can be connected to a main relief valve in parallel for convenience in adjusting for different pressure levels from one or more remote physical locations. The pilot valve with the lowest pressure adjustment will control the pressure level at the main relief valve segment which is being piloted.

Pilot Valves in Series

If the pilot valves are connected in series, the resistance to flow will be cumulative so that pressure level at the piloted main valve will be the summation of the resistance to flow through all of the pilot valves in the series.

Note the Y external drain port of Fig. 3.2(b). This can be returned

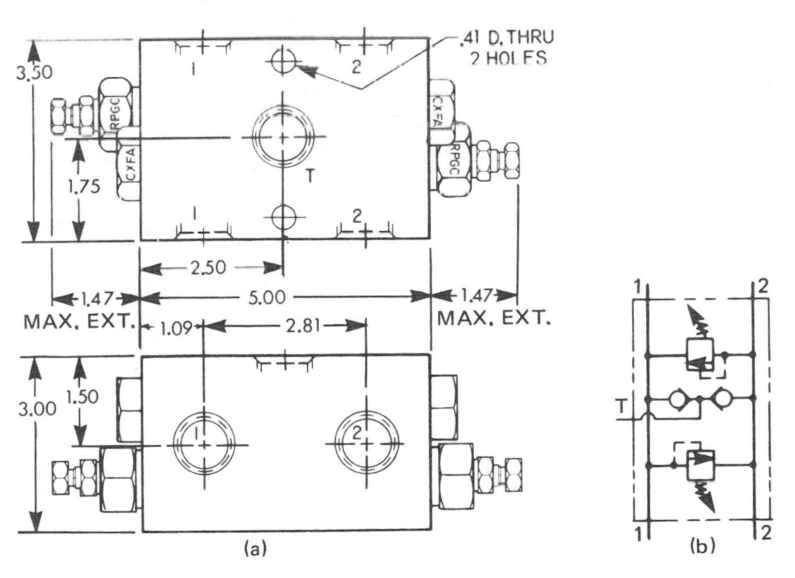

Figure 3.8 (a) Cross-port relief with anticavitation checks. (b) Symbol for cross-port relief with anticavitation checks. (Courtesy of Sun Hydraulics Corporation, Sarasota, Florida.)

unrestricted to the reservoir with minimum pressure for normal relief valve service. If a minimum pressure must be established, a second pilot valve can be connected to port Y. The integral pilot valve of Fig. 3.2(h) will determine minimum pressure at which the main valve will pass fluid. The external pilot valve (which is thus connected in series with the integral pilot valve) will establish any pressure above the minimum within the capabilities of the valve structure.

Venting of a Piloted Relief Valve

Release of the pilot fluid in the control chamber of the piloted relief valve reduces the circuit pressure at the relief valve to that created by the basic bias spring in the main piloted relief valve [Fig. 3.9(a) and (b)]. This can be accomplished by a two-way valve which can direct the pilot fluid freely to the reservoir in the desired operating pattern. A diversion valve can direct fluid from the pilot chamber of the controlled valve to several pilot valves selecting

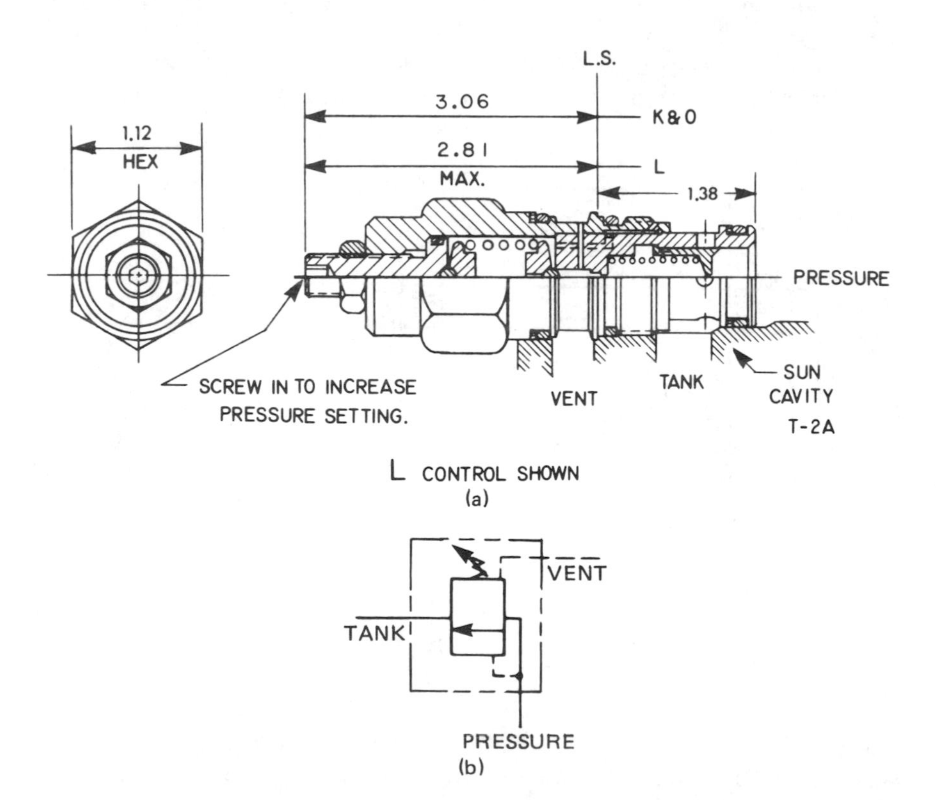

Figure 3.9 (a) Ventable relief valve cartridge. (b) Symbol for ventable relief valve cartridge. (Courtesy of Sun Hydraulics Corporation, Sarasota, Florida.)

a desired pressure level at predetermined portions of a machine circuit function.

A pilot valve can be connected in parallel to the control chamber with a secured adjustment to prevent changes of pressure beyond a predetermined maximum for safety purposes.

The key and lock adjustment shown in Fig. 3.2(b) can be used to establish maximum safe working pressure. Lesser pressure value can be established at a remote pilot valve.

Auxiliary Pilot Controls

Custom controls can be designed to establish the desired pressure value at a major relief valve assembly by electrical, pneumatic, or mechanical actuation of the pilot valve spring adjustment. The pilot need not be spring biased. It can be effected by a suitable variable orifice which is actuated by the same remote signal source.

A direct current solenoid can urge the pilot cone against the seat of the valve with area ratios that will establish a relationship between the electrical energy input which is imposed upon the solenoid winding (by a potentiometer or equivalent device) so that the pressure established by this pilot assembly will control the major poppet and thus the major circuit pressure. Certain manufacturers choose to place a mechanical spring between the end of the solenoid plunger and the cone. The solenoid effectively changes the spring loading without losing the inherent shock absorbing qualities of the mechanical spring.

If a solenoid actuated two-way valve is used to control the relief valve venting function, it can be normally open [Fig. 3.10(a) and (b)] or normally closed [Fig. 3.10(c)].

Figure 3.11(a) illustrates a subplate mounted relief valve with a solenoid operated pilot valve (14) manifolded to the pilot head assembly. Pilot pressure in channel (3) above the main poppet can be diverted through pilot valve (14) into the chamber in which valve (8) is located and then into the return line through hole (10). Figure 3.11(b) shows the external view of the valve. The solenoid pilot valve can be normally open or normally closed according to the circuit needs.

The normally-open-type limits pressure to the basic main spool bias value until the relief valve is brought up to maximum pressure potential by energizing the solenoid.

The normally-closed-type will permit maximum pressure potential until the solenoid is actuated. As the solenoid is actuated, the pilot control additive fluid is diverted to the lower pressure potential. The major relief spool responds in servo fashion, reducing the pressure to that created by the basic bias spring.

After venting or releasing the pressure assist to the major relief valve

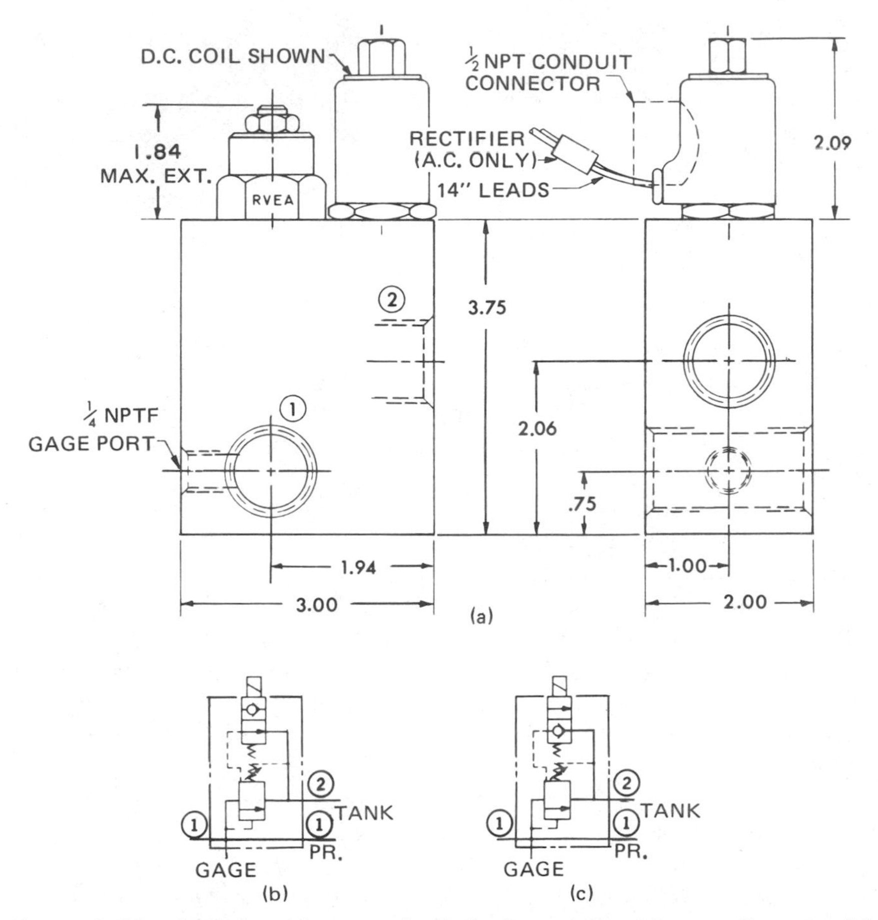

Figure 3.10 (a) Solenoid operated relief valve—solenoid normally open. (b) Symbol for solenoid operated relief valve. (c) Symbol for solenoid operated relief valve normally closed. (Courtesy of Sun Hydraulics Corporation, Sarasota, Florida.)

segment, the *rate* of reset or return to desired pressure level can be adjusted by the rate of flow back into the control chamber area [orifice 2, Fig. 3.11(a)].

Standard, off-the-shelf piloted relief valves usually have an orifice which is designed for "average" working conditions. A length of small diameter wire inserted in the orifice hole and securely fastened can slow up this reset rate. This may be desirable to reduce line shock and create a softer recovery pattern. This will not affect the release rate or safety capabilities of the major valve.

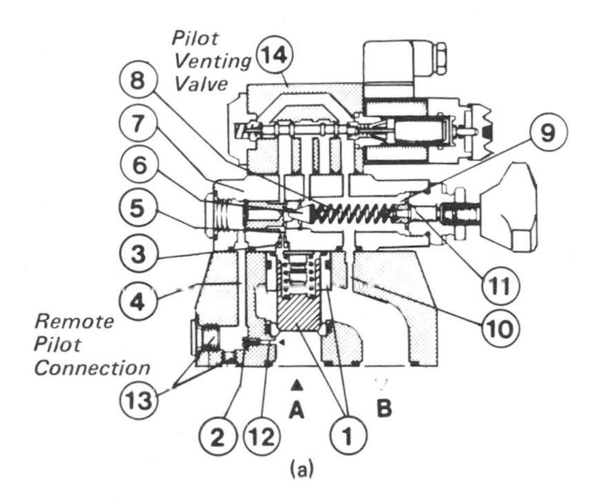

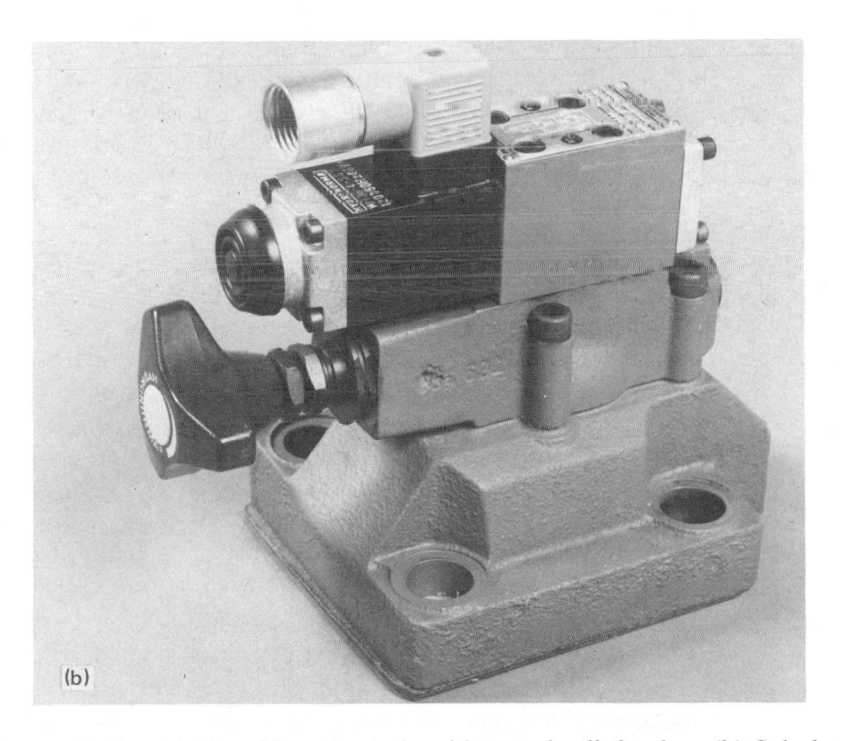

Figure 3.11 (a) Cartridge-type solenoid vented relief valve. (b) Subplate mounted solenoid vented relief valve. (Courtesy of The Rexroth Corporation, Bethlehem, Pennsylvania.)

Orifices, such as (2) in Fig. 3.11(a), may be removable so that another size can be inserted as needed.

Typical orifice diameters would be 0.023 in. or 0.031 in. Brass orifices may be machined by an electrical discharge process for precision and are usually of a sharp-edge design.

The basic piloted relief valve structure can be modified for other pressure level control functions. An accumulator of appropriate capacity can be connected in the pilot circuit to minimize pressure fluctuations and soften the overall response pattern. Many designers do every thing they can, however, to tighten the pilot circuit and raise the natural frequency.

External Drain

Several pressure levels may be needed in a hydraulic circuit. A relief valve of either the direct spring bias-type or the piloted-type can have the spring chamber connected to the reservoir by external piping. This is referred to as an external drain. The direct spring bias-type relief valve is provided with a drain to tank from the spring chamber (usually internal) so that there can be no pressure additive to the spring in usual circuit applications. A piloted relief valve can be provided with an external return to tank line from the pilot spring chamber so it, too, cannot be affected by the return fluid if unusually high pressure may be encountered in the major return to tank lines [port Y, Fig. 3.2(h)].

Rate of Pressure Rise Sensing Valves

Pressure control valves can be designed to sense rate of pressure rise. Flow is diverted to tank if pressure rise rate is excessive as would occur in a shock wave condition. The valve will stay closed if normal desired pressure rise occurs. Any pressure rise at a greater rate will cause the diversion of flow to minimize shock potential. This type of valve is commonly referred to as a shock suppressor.

3.2.5. Sequence, Priority, and Counterbalance

The major flow from a relief valve outlet port can be directed to another portion of the circuit. The input pressure will be maintained at the relief valve setting. The secondary or downstream pressure will be developed by resistance to flow and can be any pressure less than that established by the primary valve under some circuit design parameters. Of course, any drain line from the spring pocket or pilot drain must go to the tank externally. It is also possible to create a pressure greater than that set on the upstream valve. In this circumstance the upstream valve establishes a minimum pressure value in the upstream portion of the circuit and guarantees flow to the upstream area prior to flow into the downstream portion of the circuit. A valve used in this manner may be described as a *sequence* or a *priority* valve.

Integral Check Valve

The sequence or priority valve may be structured with an integral check valve for free return flow if needed in the circuit function [Fig. 3.12(a) and (b)].

A relief or safety valve will never need the return check in their basic circuit function.

Typical Applications

A typical use of a sequence valve can be found in clamping circuits. It is important that a minimum pressure be exerted on a clamping device prior to starting machining operations. The sequence valve guarantees a safe clamping pressure prior to diversion of fluid for actuation of other machine functions.

A sequence valve can also serve to insure return of work actuators and retraction of locating pins prior to unclamping. Thus, if the sequence valve is to be located in a cylinder line, it will be necessary to provide a check valve for free return flow. It is quite common to incorporate the check valve within the sequence valve structure.

Figure 3.12(c) and (d) illustrate the circuit and sequence valve assembly incorporating two pressure actuated valves and the associated check valves Flow is directed to cylinder one first. As pressure reaches a preset value and cylinder one has completed its forward stroke, fluid is directed to cylinder two. At the completion of the forward cycle the directional valve reverses the cylinder line connections. Cylinder two retracts first. At the completion of the return of cylinder two pressure builds up and passes through to cylinder one which then completes the machine cycle. The check valves are incorporated in the assembly and plumbing is at a minimum with the usual benefits of reduced potential for external leaks, lower installation costs, and simplified maintenance.

3.2.6. Counterbalance and/or Holding Valves

A counterbalance valve may be identical to a sequence valve. The major difference is in the usage and the method of draining the bias area. A sequence valve can expect to have the secondary port pressurized so it may be essential that the bias spring chamber be externally drained to tank.

Installation Options

A counter balance valve [Fig. 3.13(a)] is usually placed in a cylinder line to restrict flow and prevent movement of the cylinder member by gravity or other forces until a predetermined pressurized flow is available at designed working pressure level.

The downstream line of the counterbalance valve is usually directed to tank through the directional valve in the functional mode. Because of this

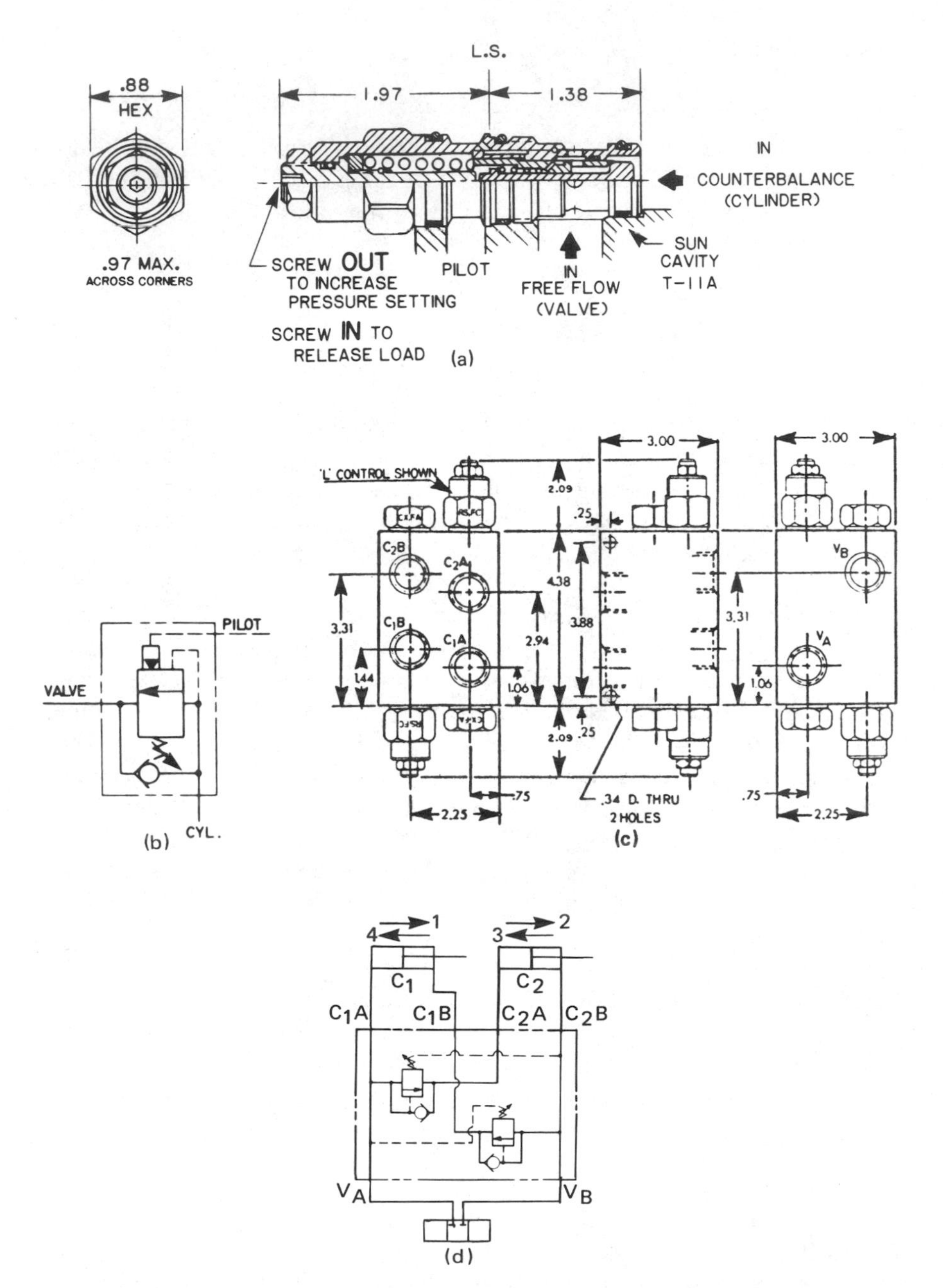

Figure 3.12 (a) Counterbalance valve cartridge. (b) Symbol for counterbalance valve. (c) Sequence valve—dual line mounted—to control the sequencing of two cylinders. (d) Symbol and circuit for sequencing two cylinders. (Courtesy of Sun Hydraulics Corporation, Sarasota, Florida.)

circumstance, the spring chamber can be drained internally. In the reverse direction of flow through the check valve, the drain is not important to the function. Thus one line can be eliminated.

Signal Source

Internal Signal Upstream The counterbalance can accept the operating signal from the valve input line. It can also be piloted from an external source. When piloted internally, it must have a pressure setting slightly higher than the load induced values.

The pressure in excess of the load pressure will determine the braking force which decelerates the load. To be most effective in many circuits it is desirable to have a *float* center position within the directional control valve. This can be with the pressure blocked and both cylinder ports connected to each other and to tank in neutral (Fig. 2.29) or all ports connected to each other and to tank in the neutral position of the directional control valve. These spool configurations permit the most effective functioning of the counterbalance and/or holding valve.

External Pilot Source By piloting the counterbalance valve from the opposite cylinder line, the external pilot source will insure a positive control of the cylinder movement. It can be used to restrain a load which tends to overrun the supply fluid from the designed source.

Counterbalance valves designed specifically to restrict cylinder travel to that resulting from the programmed pressurized flow of fluid are often identified as *holding valves*. The name probably stems from the use to hold a load to a specific movement pattern which is relatively immune to gravity conditions.

There are three specific load conditions which make a simple counterbalance valve application undesirable. They are: (1) overcenter loads, (2) varying loads, (3) and press applications where maximum tonnage is required. However, by *externally piloting* the counterbalance valve from the opposite supply line to the actuator, the overcenter counterbalance valve improves both system performance and efficiency.

The term *overcenter* comes from the fact that, in many applications, the machinery's geometry causes the load conditions to change from resistive to overrunning. This principle is represented in the example.

In the position shown in Fig. 3.13(a) you can see that the load is resisting the extension of the cylinder. If a 3000 lb. force is required at point A to move the load, and the cylinder has a piston area of one square inch, then 3,000 psi pressure is created in the fluid being supplied by the pump. At any pressure over 100 to 200 psi, the overcenter counterbalance valve is piloted wide open, offering little resistance to flow from the rod end of the cylinder.

After the load moves past the center line [Fig. 3.13(b)], the geometry

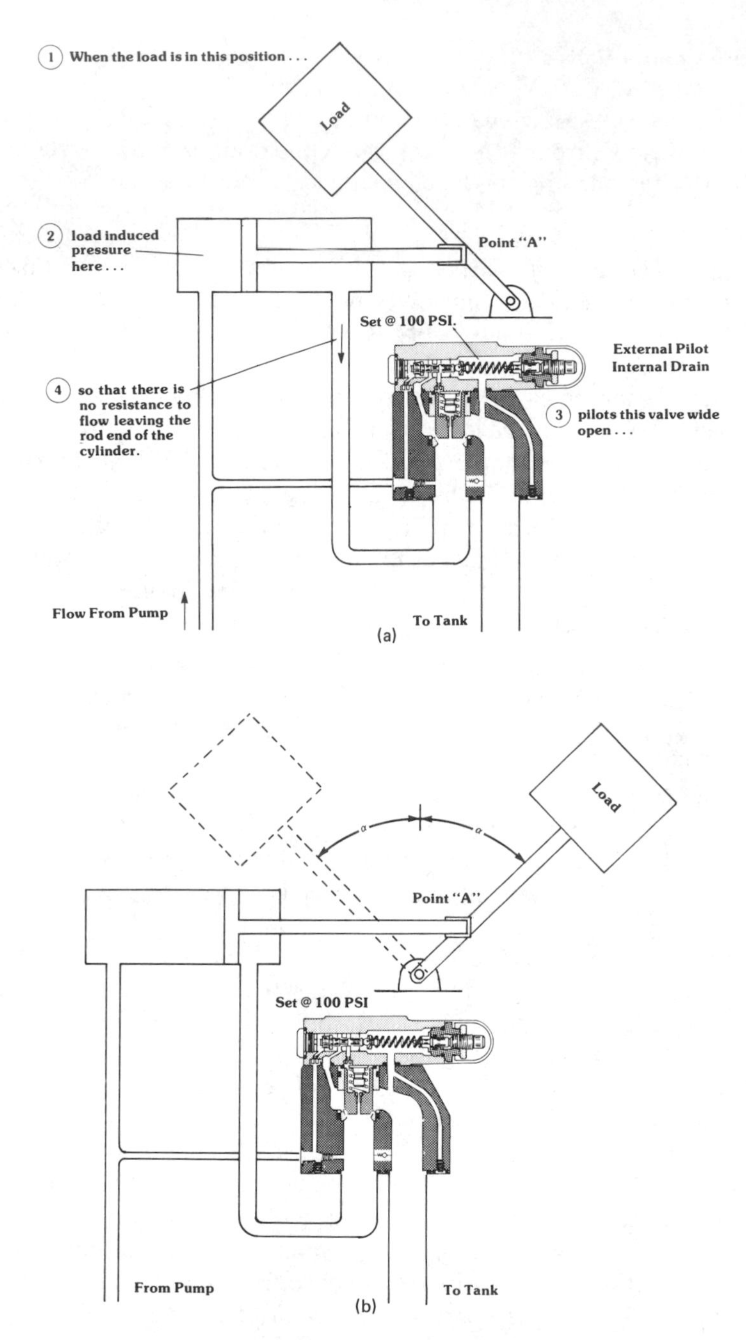

Figure 3.13 (a) Circuit for resistive load. (b) An overrunning load. (Courtesy of The Rexroth Corporation, Bethlehem, Pennsylvania.)

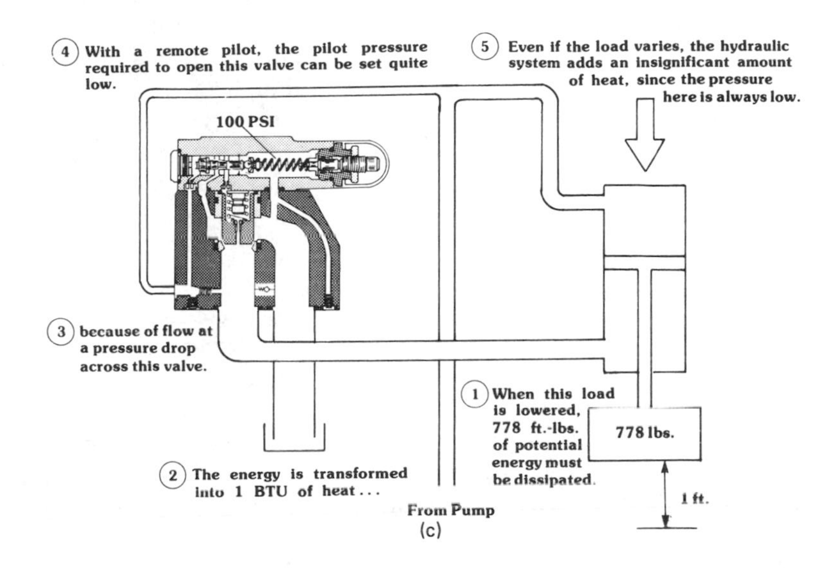

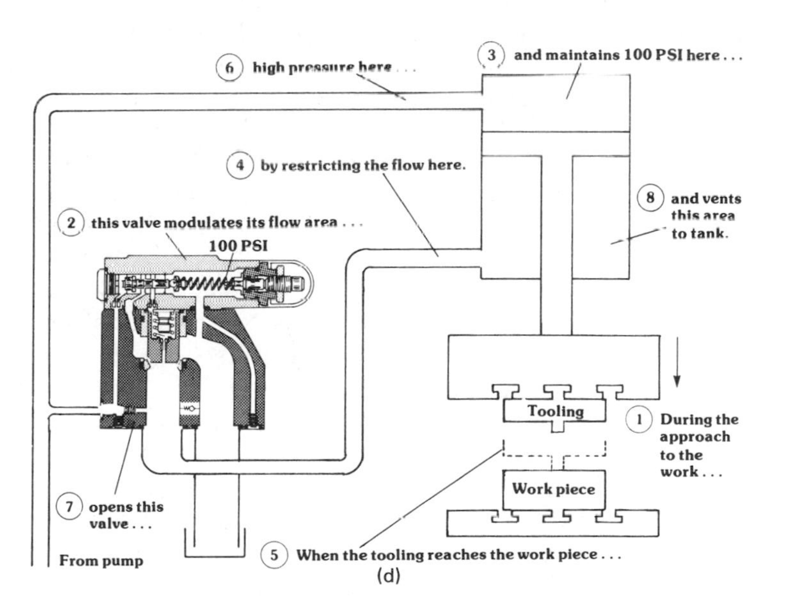

Figure 3.13 *(continued)* (c) External pilot for counterbalance valve. (d) Counterbalance circuit for maximum press tonnage. (Courtesy of The Rexroth Corporation, Bethlehem, Pennsylvania.)

of the load forces now pulls on the cylinder rod. When the load is at the same angle on the other side of center, a 3000 lb. load force is trying to extend the cylinder. Since the effective area on the rod end of the cylinder is slightly less than one square inch, a pressure slightly higher than 3000 psi is needed to keep the load from free falling.

Because the overcenter type counterbalance valve is externally piloted, its opening and closing is not affected by pressure conditions in the rod end of the cylinder. It only responds to a pressure signal in its external pilot line, or, in this case, in the blind end of the cylinder.

When the load goes overcenter, the load forces try to pull the cylinder's piston ahead of the oil supply from the pump. Under these conditions, there is no resistance to flow from the pump. In fact, the moment the piston gets ahead of the oil stream, a vacuum condition is created on the blind end of the cylinder.

Because of the pressure override characteristics of the counterbalance valve, a change in pressure is necessary to fully open the valve. In our example, if the valve is set at 100 psi cracking pressure, the full open position does not occur until a pilot pressure of 200 psi is reached.

As the load moves overcenter, pressure is lost in the blind end of the cylinder. Consequently, a tendency for the closing of the counterbalance valve exists. The counterbalance valve resists the motion of the load with whatever pressure is necessary to maintain a 100 to 200 psi pressure in the fluid being supplied by the pump.

You will notice that we have explained the operation of our overcenter counterbalance valve *in only one direction of operation*. For reverse motion of the load, *a similar valve would have to be used on the blind end of the cylinder*.

Overcenter loads can encounter varying load conditions. The major advantage of using an externally piloted overcenter type counterbalance valve, with varying load, is that is greatly improves system efficiency. We said earlier, in our discussion of the simple counterbalance valve, that the pressure setting of the valve should be set slightly higher than the maximum load induced pressure. We also should have mentioned that any reduction in load meant that a higher pressure had to be supplied to the opposite end of the actuator to compensate for the loss in load induced pressure. In turn, this unnecessarily high pressure is converted into heat as flow is forced across the counterbalance valve.

When the externally piloted counterbalance is used, the valve becomes insensitive to the load induced pressure. The valve allows movement of the overhung load as long as a minimum pressure (approximately 100 psi, depending on the valve setting) is maintained in the opposite end of the actuator. You can see that, under these conditions, the hydraulic system never substantially adds to the heat generated in lowering the load [Fig. 3.13(c)]. Of course, there is no way to prevent the potential energy of the load from being converted into heat, since the energy must be expelled as the load lowers. This must be taken into consideration when you are calculating the heat generated in the hydraulic system.

Figure 3.13(d) shows how the external pilot of the counterbalance valve limits the energy loss as the press closes ready for maximum tonnage on the die.

Thermal Relief Considerations The holding valve assembly may also include a thermal relief structure and a check valve for free flow return of the fluid.

Differential Pilot Signal Some models can be fitted for different pilot pressure ratios appropriate to the circuit requirements.

With Shuttle Valve A shuttle valve network may be designed into the pilot circuitry to accept the highest source of pilot pressure in an automatic mode.

Safety Considerations

Holding valves may be manufactured in cartridge form so that they can be fitted into the cylinder cap housings to eliminate connections and potential loss of control resulting from a ruptured hose or pipe line (Fig. 3.12). Some models are manifolded to the cylinder for the same reason. Holding valves are always internally drained.

3.2.7. Unloading Valves

Unloading valves are normally closed, two-way valves that are externally piloted and often internally drained (Fig. 3.14).

Pump Unloading

The purpose is to divert fluid to a secondary port at minimum resistance to flow. Most common usage is to selectively unload one or more pumps to the reservoir when a source of pilot pressure is available. The action can be automatic from a pressure rise signal or through some other signal source.

Figure 3.15(a) shows a valve assembly consisting of a high-pressure relief valve (P_1 – high pressure), an integral check valve allowing flow from P2 to

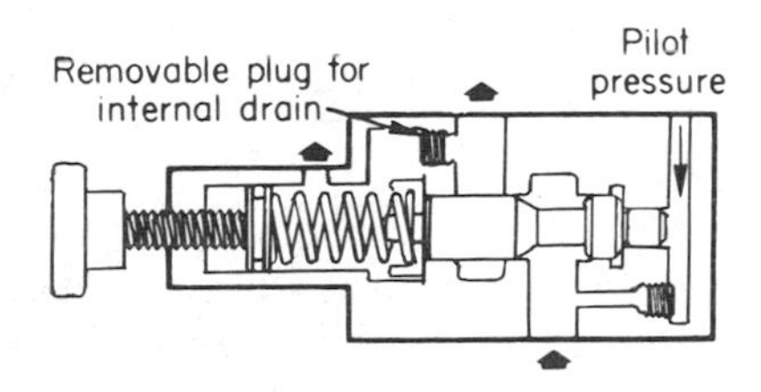

Figure 3.14 Unloading valve, externally piloted, externally drained. (Courtesy of Double A Products Company, Manchester, Michigan.)

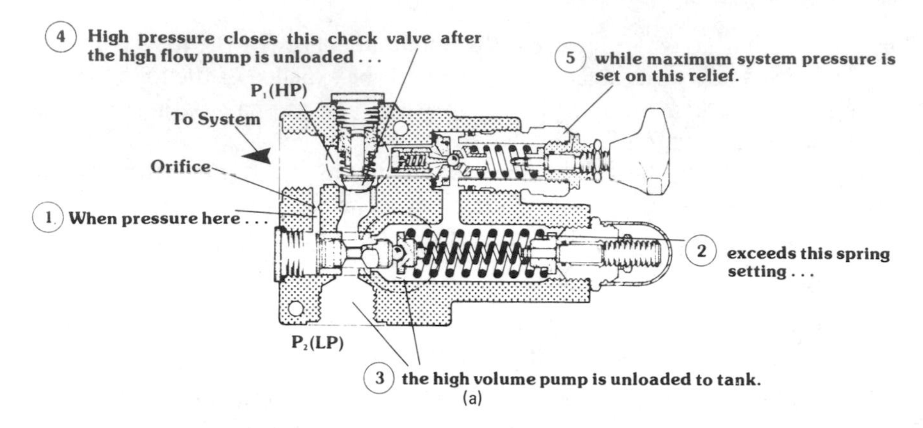

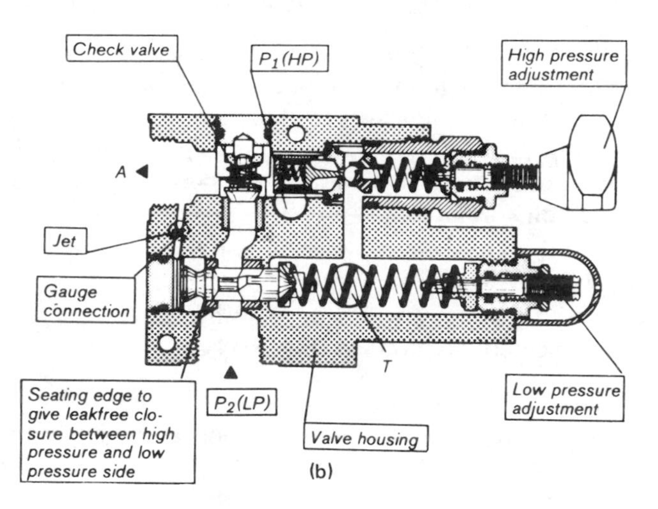

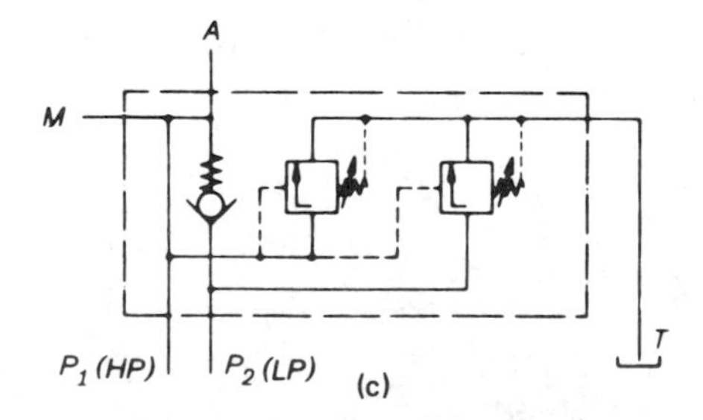

Figure 3.15 (a) Relief, unloading, and check combination. (Courtesy of The Rexroth Corporation, Bethlehem, Pennsylvania.) (b) Unloading valve circuit. (c) Symbol.

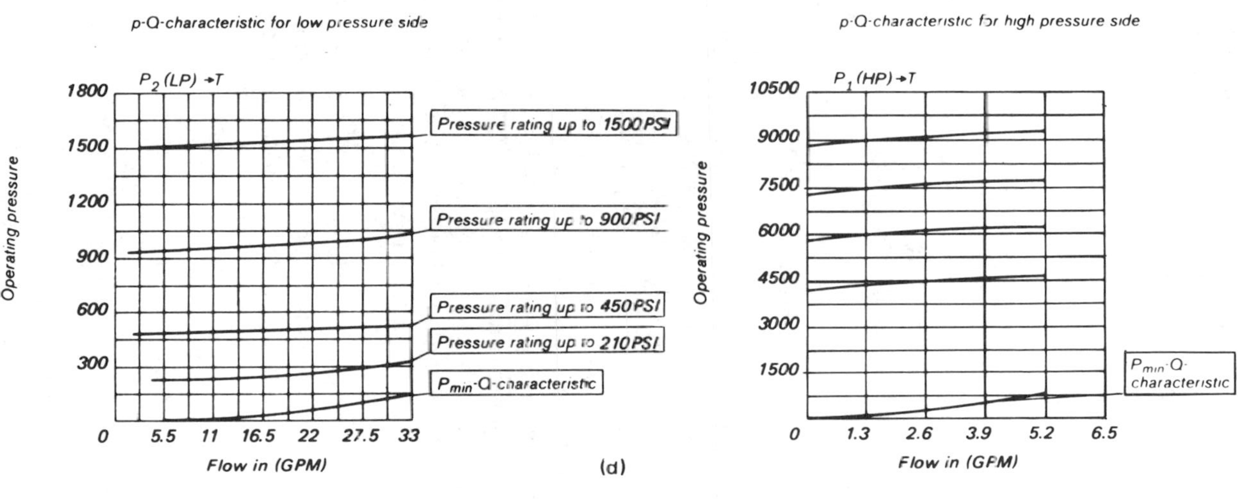

Figure 3.15 *(continued)* (d) Typical operating characteristics for unloading, relief, and check valve combination. (Courtesy of The Rexroth Corporation, Bethlehem, Pennsylvania.)

join with flow from P1 at low pressure, and an unloading valve to divert flow from P2 to tank at a predetermined pressure level. Schematic and parts identification are shown in Figure 3.15(b,c). Figure 3.15(d) illustrates typical flow curves for the low-pressure pump prior to unloading and the high-pressure curves for the control pump.

Unloading of Pumps in an Accumulator Circuit

An unloading valve designed with a pilot assembly structured to accept a pressure signal from an accumulator is used to divert pump flow to tank when the accumulator is fully charged. At 10% to 25% below full charge pressure, the valve will close, causing the pump flow to pass through a check valve to recharge the accumulator and/or to supply working fluid to the circuit in conjunction with the flow from the accumulator. The pilot signal is taken between the outlet of the check and the input of the accumulator (Fig. 3.16).

All accumulator circuits must be provided with some means of releasing the pressure in the circuit as dictated by the machine operating characteristics.

Certain codes mandate that an automatic release mechanism be incorporated in the circuit so that it will function and release pressurized fluid from the accumulator when the machine is shut down for any purpose. The automatic release mechanism can be an electrically, pilot, or manually operated valve. Special flow sensing valves can be structured to release the accumulator charge when flow from the pump stops.

The accumulator unloading valve illustrated in Fig. 3.17 has been designed specifically for use in accumulator circuits. Its' design provides three functions. It limits maximum system pressure, unloads the pump to tank when the accumulator reaches the desired pressure, and it reloads the pump to bring the accumulator up to full charge after a predetermined minimum pressure has been reached.

An accumulator stores a given volume of fluid under pressure. At a relatively low pressure, the accumulator stores less fluid than it does at full system pressure. Actually, the fluid volume available from an accumulator

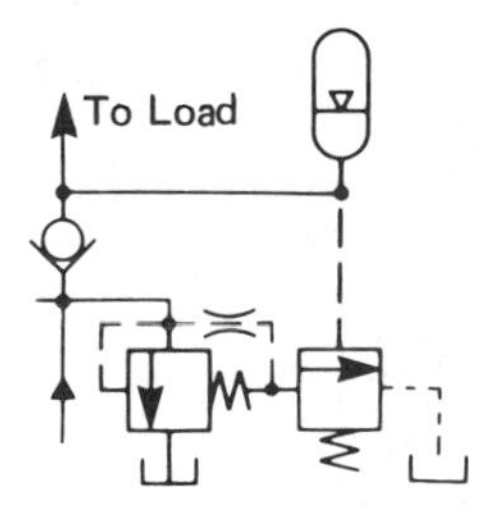

Figure 3.16 Typical accumulator unloading circuit.

Figure 3.17 Accumulator unloading valve. (Courtesy of The Rexroth Corporation, Bethlehem, Pennsylvania.)

is determined by subtracting the fluid volume held at minimum operating pressure from the quantity of fluid stored at maximum system pressure. This differential volume is discharged by the accumulator as system pressure drops from maximum to minimum. Of course, the time period during which the drop in pressure occurs establishes the flow rate available from the accumulator. Remember:

$$\text{Flow} = \frac{\text{volume}}{\text{time}}$$

When used with a fixed displacement pump, the accumulator unloading valve performs both a relief and an unloading function. The valve consists of a cartridge poppet pilot-operated relief valve, an isolating full flow check valve, and an unloading piston, which overrides the pilot relief function. Let us now consider the various operating conditions of this valve.

The operational cross section [Fig. 3.18(a)] shows that this valve has three working ports: a pump inlet, a system connection, and a tank return.

The pump's outlet is connected directly to the pressure port of this valve, so that all the pump flow must pass through the valve, before entering the system. Of course, the tank port is connected directly to the reservoir. If your choice is to route the return flow through filters or heat exchangers, you should use an external drain for the spring chamber of the pilot.

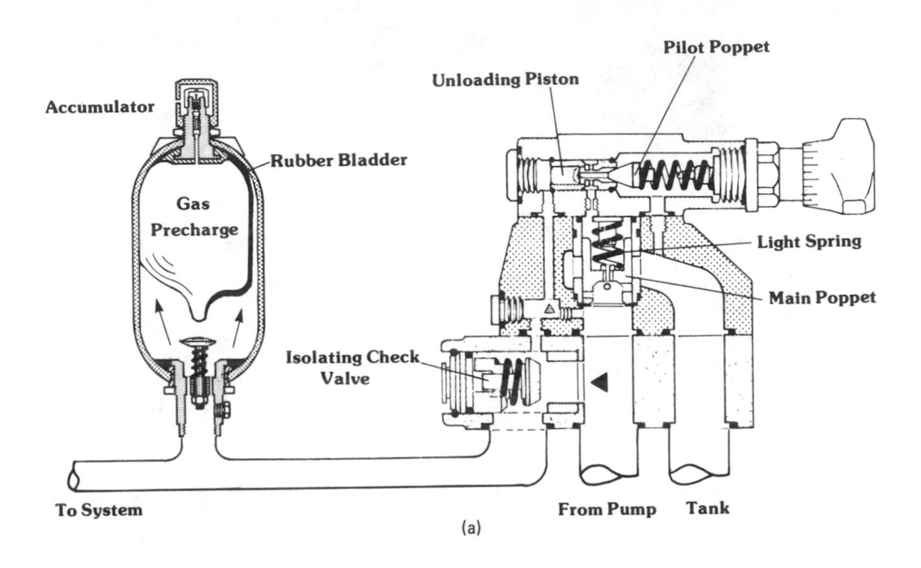

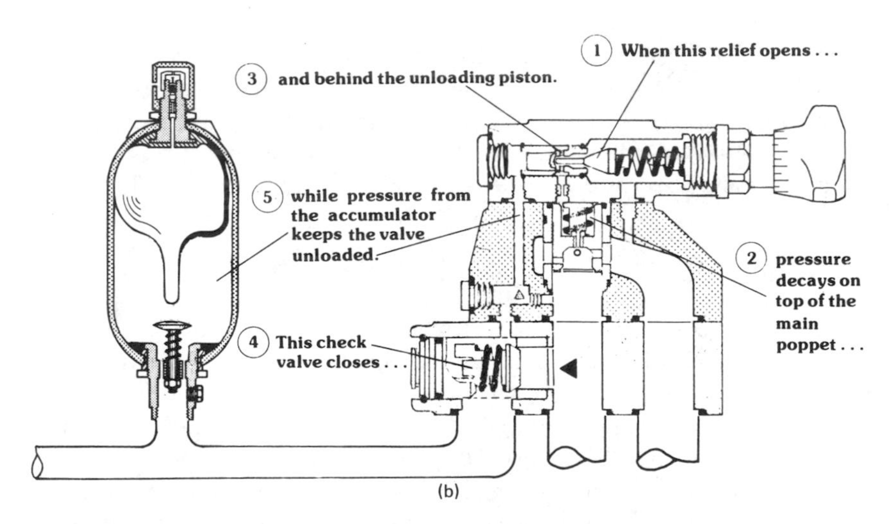

Figure 3.18 (a) Charging the accumulator. (b) Relieving and unloading of the pump. (Courtesy of The Rexroth Corporation, Bethlehem, Pennsylvania.)

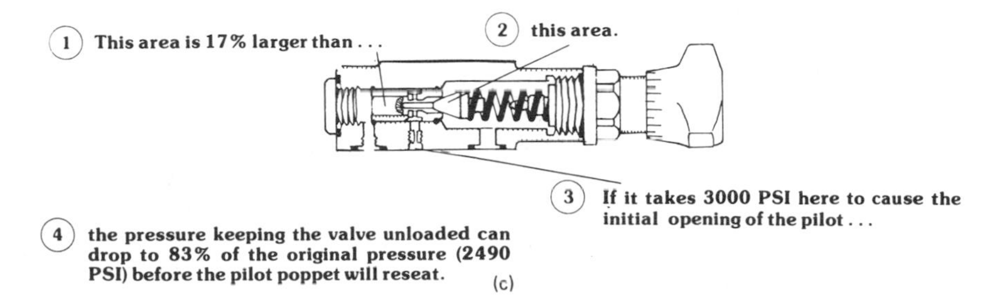

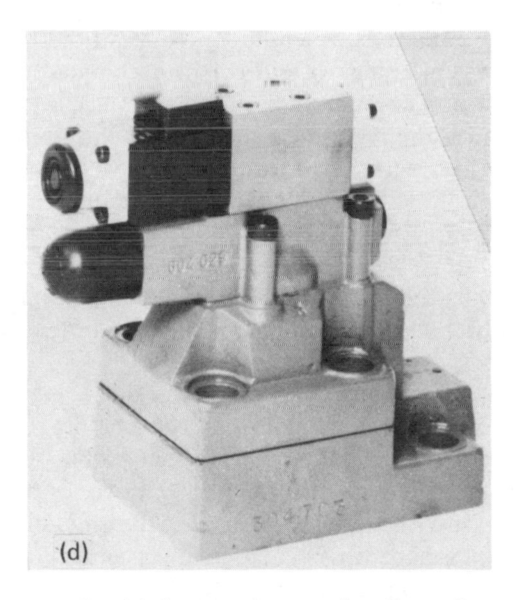

Figure 3.18 *(continued)* (c) Accumulator unloading pilot section. (d) Solenoid controlled accumulator unloading valve. (Courtesy of The Rexroth Corporation, Bethlehem, Pennsylvania.)

Assuming that the pilot spring seats the pilot poppet, the main poppet is also closed, due to the hydraulic pressure balance and the light spring force. The valve delivers flow to the system over the isolating check valve, which in turn charges the accumulator as pressure in the system increases. As long as there is flow, a higher pressure exists on the right-hand area of the unloading piston than on the left due to the pressure drop across the isolating check valve. The unloading piston is held in the left-most position, and has no effect on the relieving function.

When the accumulator has reached its desired charge, system pressure unseats the pilot relief, which causes a decay in system pressure above the main poppet [Figure 3.18b)]. At the same time, pressure is lost on the right-hand area of the unloading piston, so that pressure in the system from the accumulator holds the unloading piston against the nose of the pilot poppet, keeping it unseated. The moment the main poppet opens, a pressure loss at the inlet of the check valve causes it to close, thus isolating the pump from the rest of the system.

Under these conditons, the pump circulates fluid freely to tank while pressure in the system is maintained by the accumulator.

The purpose of an accumulator unloading valve is to prevent the pump from being reloaded the moment a slight decay in system pressure occurs. This is exactly the reason a standard unloading valve cannot be used in an accumulator application. That is, with a standard unloading valve, the small pressure differential between open and closed positions will set up a rapid cycling of the pump between the loaded and unloaded conditions. To overcome this problem, the pilot head of an accumulator unloading valve is designed with differential effective areas between the pilot relief and the unloading piston.

Figure 3.18(c) shows how the accumulator unloading valve allows system pressure to fall a predetermined minimum value before reloading the pump. Since the effective area of the pilot poppet is smaller than that of the unloading piston, a higher pressure is needed to initially move the pilot poppet against the spring force. Once the right-hand area of the unloading piston is vented over the opened pilot poppet, system pressure becomes effective on the larger pilot area of the unloading piston. Of course, the larger area means more available force, so that we can keep the pilot poppet unseated with a somewhat lower pressure. The area ratio is usually in the neighborhood of 17% for the units shown in these illustrations.

Precautions: The function of the accumulator unloading valve should not be confused with that of an accumulator safety valve. Once the accumulator is charged, the unloading valve has no means of bleeding the accumulator charge to tank if the system is shut down. Likewise, a fully charged accumulator is not protected from overpressure due to increased load or thermal expansion. The accumulator unloading valve only provides pressure protection in regard to the pump's capability to pressurize the system.

Like the solenoid venting feature for pilot-operated reliefs, a directional control valve can be added to the pilot section of the accumulator charging valve [Fig. 3.18(d)]. An electrical signal can override the unloading of the pump, as was already shown in the discussion of pilot-operated reliefs.

3.2.8. Reducing Valves

All of the preceding pressure control valves have been *normally-closed* two-way valves which are piloted from an *upstream* source or from an *external* pilot source.

A reducing valve is a *normally-open* two-way valve responding to a signal from the *downstream* source (Fig. 3.19).

The reducing valve components of Fig. 3.20 are fitted into a housing which is subplate-mounted to the plate in which the connecting pipes terminate. Note the basic cartridge structure which can be housed in any appropriate envelope.

Pressurized liquid entering at B is free to pass through to A. Spool (1) is provided with a restricted passage (2) into the bias spring chamber. Orifice (3) is connected to the bottom of the spool which opposes the bias spring. Orifice (4) is parallel to but longer than orifice (3). Poppet (5) is seated in the upper end of passage (7) and is held in position by spring (8). Drain fluid passes through channel (9) to the Y port or to tank. A pilot valve can be connected in series at port Y. Check valve (10) can be replaced with a plug device if a return flow is not needed.

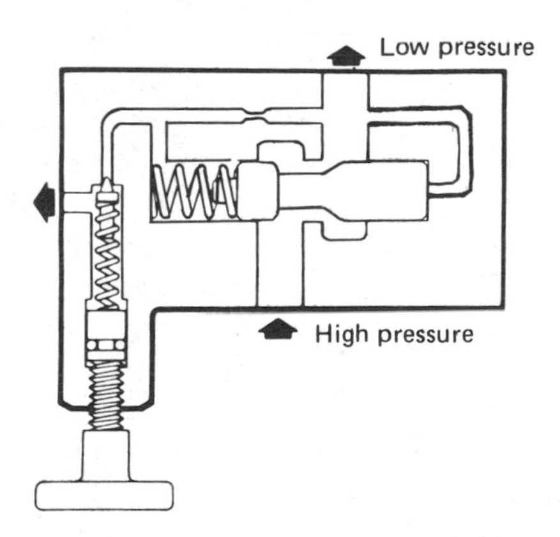

Figure 3.19 Reducing valve. (Courtesy of Double A Products Company, Manchester, Michigan.)

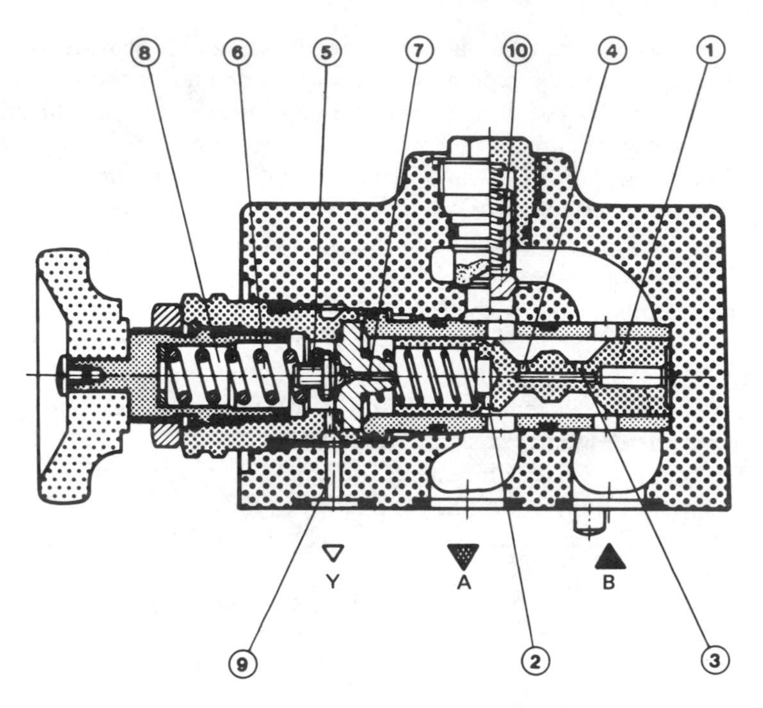

Figure 3.20 Subplate mounted cartridge reducing valve. (Courtesy of The Rexroth Corporation, Bethlehem, Pennsylvania.)

In operation the pressure level at A is sensed through the orifices at (3) and (4) to the bottom of the spool. At a predetermined pressure level, flow past poppet (5) will be greater than flow through orifice (2), creating an unbalance of spool (1) urging spool (1) to the left and restricting major flow from B to A. A pressure balance will be established by poppet (5) and spring (8) such that spool (1) will modulate and maintain the desired pressure at A.

A structure such as that shown in Fig. 3.21(a) may be needed when large flows and major pressure fluctuation are encountered in the reducing valve action. The valve of Fig. 3.21(a) senses downstream pressure in the conventional manner and modulates the normally-open poppet in sleeve (1) much like the valve shown in Fig. 3.20. The major difference is that pressure to assist the bias spring on the major spool structure is taken from the input B port and passed through a compensating flow control valve [Fig. 3.21(b)] before entering the control chamber through passage (4). Poppet (6) determines maximum pressure assist and resulting pressure level at A.

In the static, no flow condition of the valve (main spool fully closed), an overload at the actuator, or leakage oil would tend to increase pressure

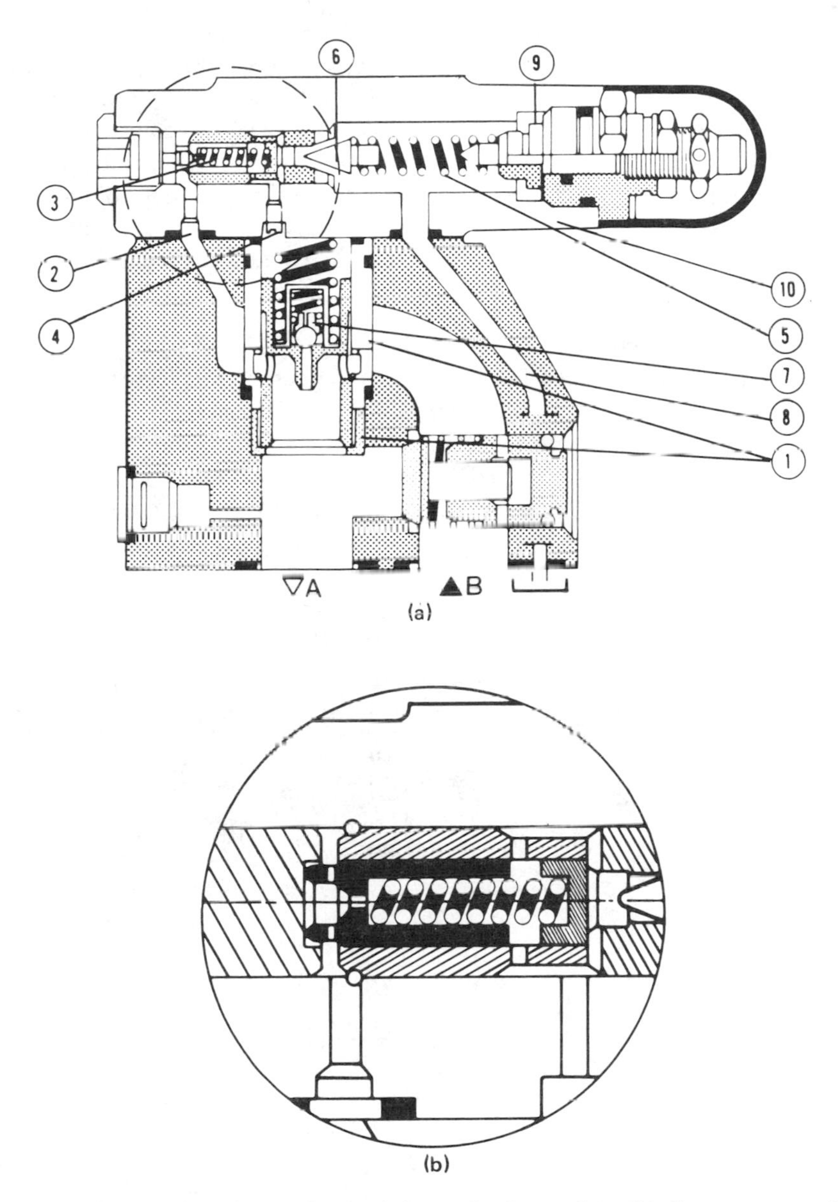

Figure 3.21 (a) Primary control for reducing valve. (b) Compensator for primary control reducing valve. (Courtesy of The Rexroth Corporation, Bethlehem, Pennsylvania.)

in secondary (Port A). Relief protection is provided through the pilot section by a miniature relief valve poppet (7) built into the main control spool (1). The pilot section (10) must always be externally drained through the Y port (8 or 9).

It is somewhat obvious that the designation *reducing* valve can be misleading. The function may not be simply that of physically reducing pressure from one level to another. The actual function is to sense and control the pressure level beyond the normally-open two-way valve structure and reduce or close off the supply to the downstream area when a desired pressure level has been attained as preset on the adjusting mechanism.

Protection from Overpressurization

Should external forces cause the pressure to increase in this secondary area, the conventional valve can do nothing to control the increase. Some valve designs [Fig. 3.19, 3.20, 3.21(a), 3.22(a), and 3.22(b)]) include a relief function and pilot control which serves to eliminate pressure increase by letting some fluid return to tank. The complete control function can be accomplished with one single adjustment.

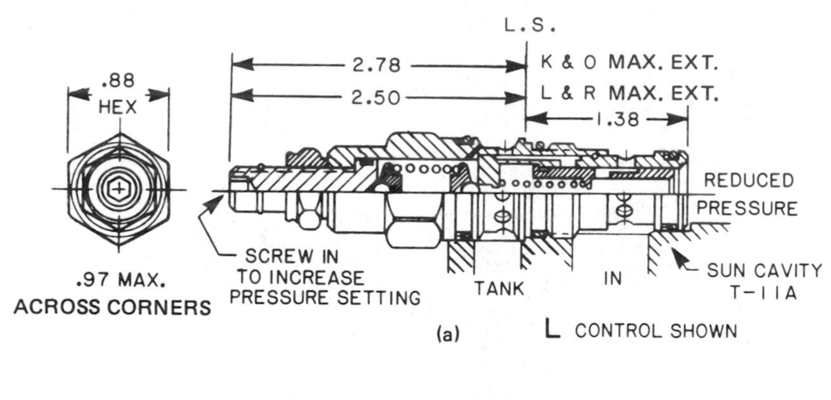

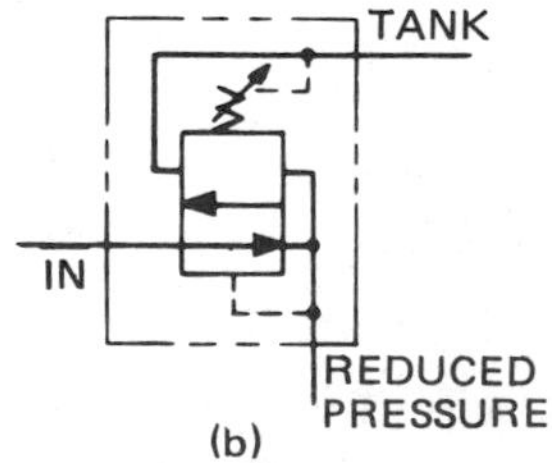

Figure 3.22 (a)Reducing valve cartridge. (b) Symbol for reducing valve. (Courtesy of Sun Hydraulics Corporation, Sarasota, Florida.)

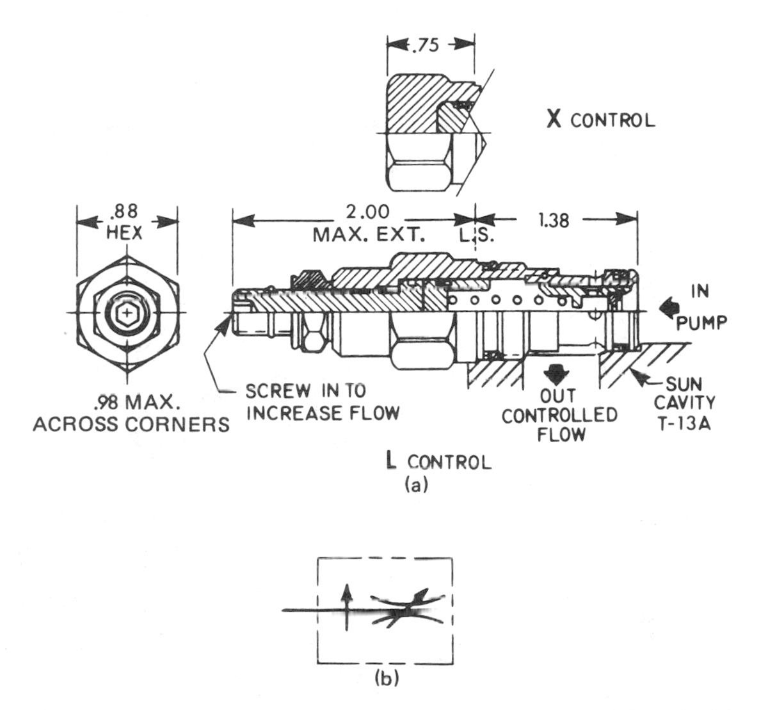

Figure 3.23 (a) Pressure compensated flow regulator. (b) Symbol for flow regulator. (Courtesy of Sun Hydraulics Corporation, Sarasota, Florida.)

Pressure Compensation for Flow Valves

Reducing valves are used for maintaining a uniform flow across an orifice. Many flow control mechanisms contain a reducing valve used to insure pressure across the orifice or act as a pressure compensator [Fig. 3.23(a) and (b)].

3.2.9. Pressure Switch

A pressure switch is a device to interface between a fluid circuit and an electrical circuit.

A rising or dropping pressure can operate an electric switch. The purpose can be to provide a simple signal to indicate pressure level or to actuate electrically-operated valves for various machine functions responsive to pressure variables.

The pressure switch is a digital off-on device. Figure 3.24(a) illustrates a typical piston-type pressure switch. Similar switches can be actuated by a Bourdon-tube structure. Technical data for a typical switch is given in Fig. 3.24(b). The associated electrical data is shown in Fig. 3.24(c). Symbols are

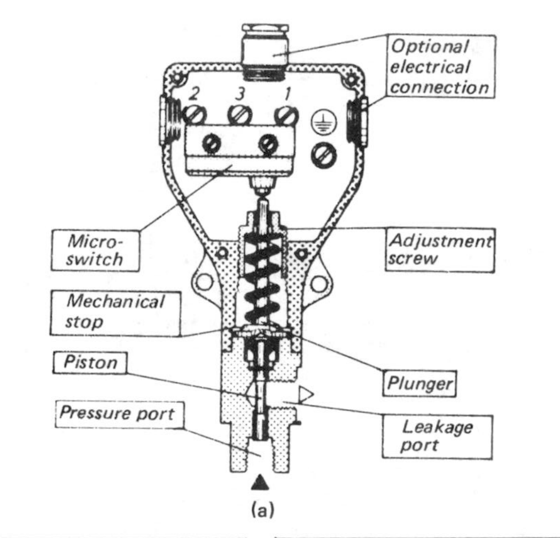

(a)

Switching accuracy	< ± 2% of pressure setting
Max. switching frequency	HED 1 KA (300/min.)
	HED 1 OA (50/min.) (briefly also 100/min.)
Electrical conduit connection	1/2" NPT
Hydraulic medium	Mineral oil (please consult us for other fluids)
Fluid temperature range	−4 to +160° F
Viscosity range	35 to 1750 SSU
Mounting position	Optional

Pressure setting range in PSI				Type	Max. working pressure PSI (briefly)
Falling pressure		Rising pressure		HED 1 K . . .	
min.	max.	min.	max.		
45	1380	90	1500	HED 1 KA 20/100	9000
90	4875	150	5250	HED 1 KA 20/350	9000
150	6975	300	7500	HED 1 KA 20/500	9000

Max. pressure at leakage connection		30 PSI		
Insulation: DIN 40 050		IP 65		
Contact loading	AC Voltage		DC Voltage	
	Volts	Amps	Volts	Amps
	460	15	40 125 250	1,0 0,4 0,2
With higher DC voltages arc suppression is recommended to increase working life.				

Pressure setting range in PSI				Type	Max. working pressure (PSI) (briefly)
Falling pressure		Rising pressure		HED 1 O . . .	
min.	max.	min.	max.		
30	675	53	750	HED 1 OA 20/ 50	750
45	1230	120	1500	HED 1 OA 20/100	5250
90	4425	300	5250	HED 1 OA 20/350	5250

(b)

Figure 3.24 (a) Piston-type pressure switch. (b) Technical data for pressure switch. (Courtesy of The Rexroth Corporation, Bethlehem, Pennsylvania.)

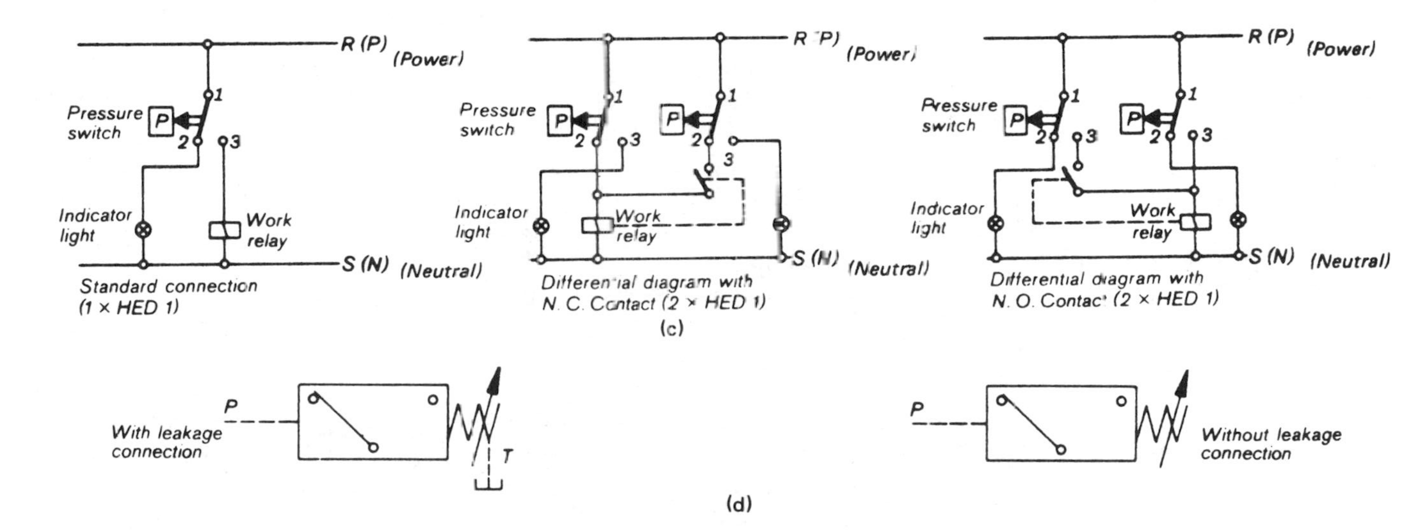

Figure 3.24 *(continued)* (c) Electrical diagram for pressure switch. (d) Pressure switch symbols. (Courtesy of The Rexroth corporation, Bethlehem, Pennsylvania.)

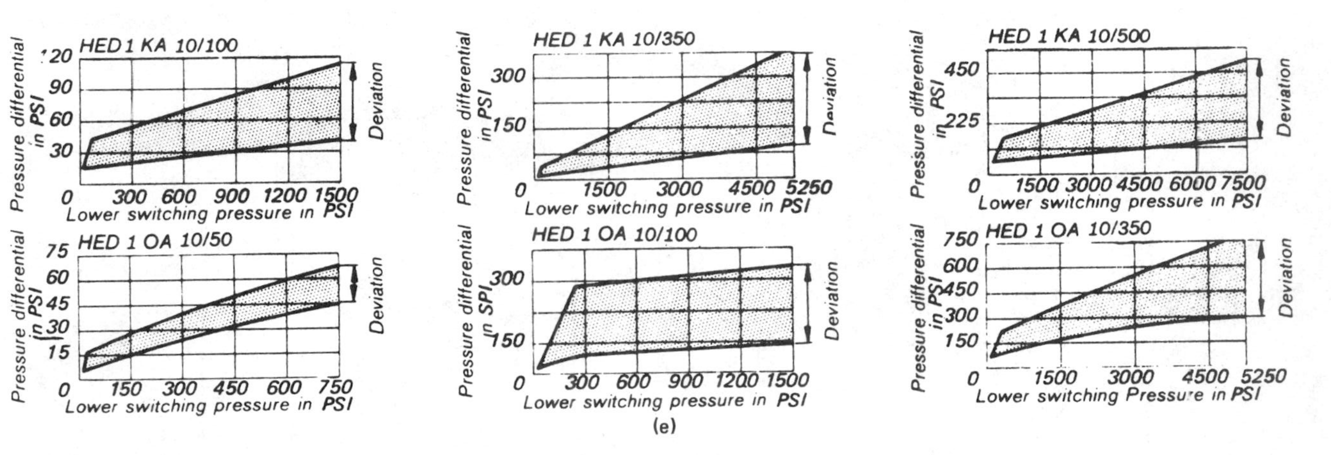

Figure 3.24 *(continued)* (e) Pressure differential data. (Courtesy of The Rexroth Corporation, Bethlehem, Pennsylvania.)

shown in Fig. 3.24(d) and pressure differential characteristics are given in Fig. 3.24(e).

3.2.10. Pressure Transducer

A pressure transducer gives a random resistance related to the pressure level which can supply an irregular electrical signal to be amplified if needed for use as a control for a variable force motor or provide a readout indication to remotely monitor pressure level.

3.3. PRESSURE CONTROL VALVE COMBINATIONS AND SPECIAL PURPOSE ADAPTATIONS

3.3.1. Relief Valve Adaptations

Flanged to Pump Outlet

Minimizing plumbing, especially in the larger pipe sizes, can reduce first cost and minimize flow losses. The valve of Fig. 3.25(a) and (b) is designed to sandwich between the outlet pipe flange and the pump body.

Directional Control Valve Sandwich

Machine tool-type directional control valves can be manifolded to a subplate in which the piping terminates. The relief valve can be inserted in a manifold plate sandwich between the subplate and the directional valve or physically located in the subplate as shown in Fig. 3.25(c), and by symbols in Fig. 3.25(d). Several modifications of these plates are available so that relief from port to port can be arranged appropriate to the circuit needs.

3.3.2. Dual Counterbalance Valves With Cross-Pilot Assist

Many circuit applications require a counterbalance function in each direction of travel. A valve such as Fig. 3.26(a) and (b) combine the two functions in a single housing. The integral check permits independent operation and adjustment in each direction.

With Built-In Shuttle Valve

The shuttle valve (Fig. 3.27) integrated into the dual counterbalance valve provides a pressure source for hydraulically releasing a spring applied mechanical brake or as a load pressure signal for load sensing circuits.

A single counterbalance valve can also be structured with an integral shuttle valve as shown in Fig. 3.28(a) and (b).

3.3.3. Regenerative Valving

Rod End Control With Integral Check Valves

Regenerative circuits allow a cylinder to be advanced more rapidly than it could be with pump flow alone. To achieve this, oil from the rod end of a

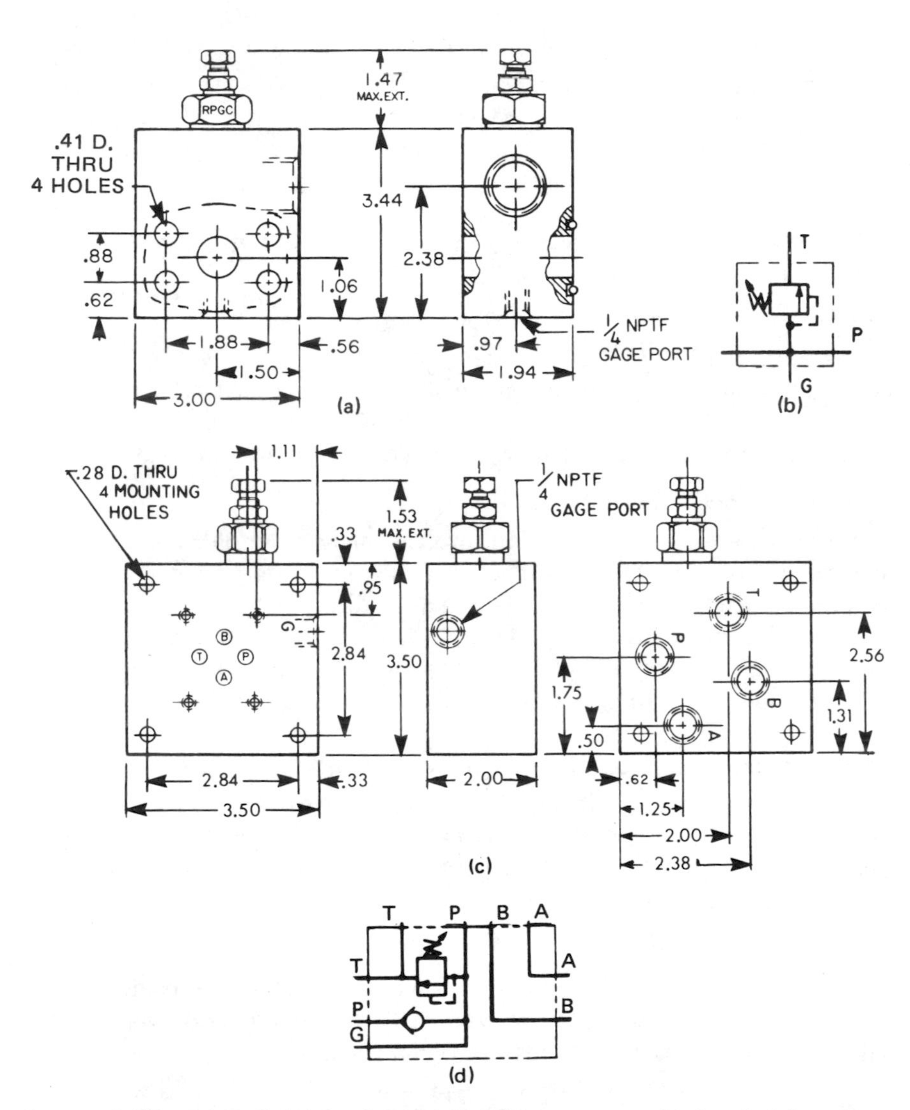

Figure 3.25 (a) Relief valve for manifolding to pump. (b) Symbol for pump mounted relief valve. (c) Relief valve in subplate. (d) Symbol for relief valve in subplate. (Courtesy of Sun Hydraulics Corporation, Sarasota, Florida.)

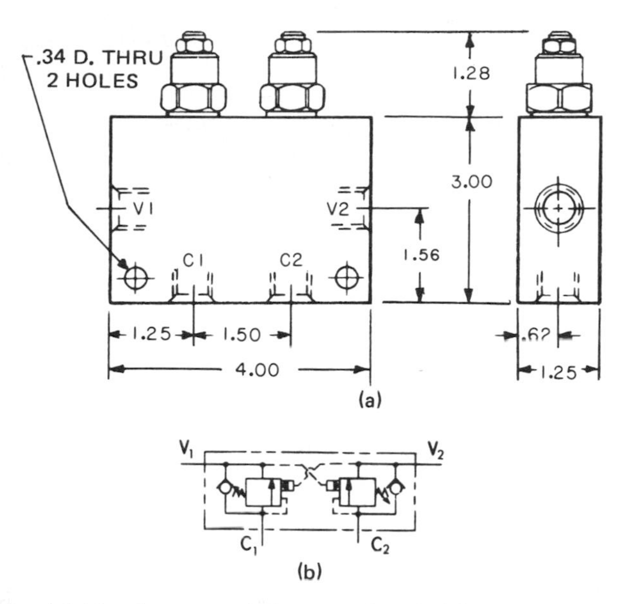

Figure 3.26 (a) Dual counterbalance valve (with cross pilot assist). (b) Symbol for dual counterbalance valve. (Courtesy of Sun Hydraulics Corporation, Sarasota, Florida.)

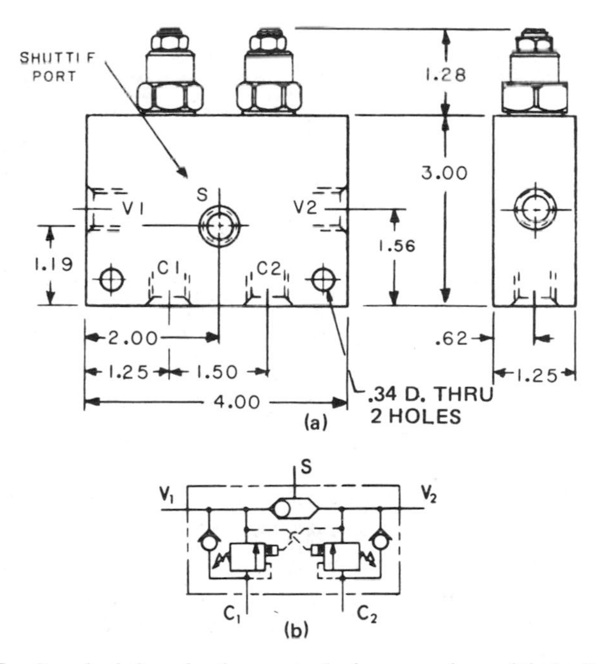

Figure 3.27 Symbol for dual counterbalance valve with built in shuttle. (Courtesy of Sun Hydraulics Corporation, Sarasota, Florida.)

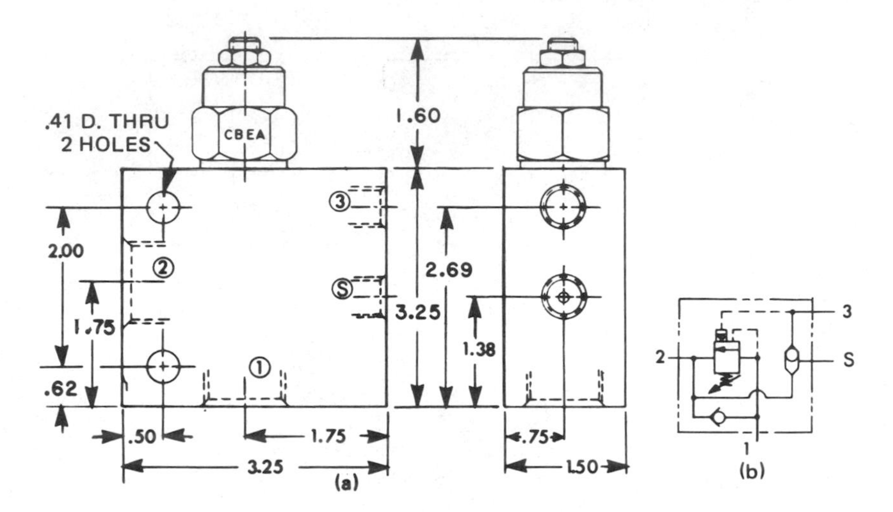

Figure 3.28 (a) Counterbalance valve with pilot assist and built-in shuttle valve. (b) Symbol for counterbalance with pilot assist and shuttle valve. (Courtesy of Sun Hydraulics Corporation, Sarasota, Florida.)

cylinder is added to the pump flow to the blind end, increasing the rate of advance [Fig. 3.29(a) and (b)].

Regenerative valve directs rod end oil to join with the pump flow to the blind end for rapid advance during the regenerative part of the cycle. When the work is engaged and pressure in the blind end rises to the setting of the regenerative valve, rod end oil is automatically diverted to the reservoir. The circuit then becomes nonregenerative and full system pressure is applied to the blind end, allowing it to develop maximum force.

In the cylinder return mode, pressurized fluid is directed through a check valve to the rod end of the cylinder. Pilot pressure is applied to the check valve used to direct fluid from the rod end to the head end so that fluid used in the retraction mode is not lost through this valve at this time in the cycle.

The sequence valve used in this assembly uses a cross-pilot assist. As the pressure approaches the set value, regenerative flow decreases progressively.

Cylinders are commercially catalogued with rod area roughly one-half head area so that the regenerative speed is approximately double that of usual mode.

The valves illustrated in Fig. 3.29(a) and (b) are designed so that only regenerative (rod end) oil—or pump flow during retraction—passes through these valves.

Regenerative Circuit with Separate Off-the-Shelf Valving

The circuit of Fig. 3.30 requires a relationship that entails the use of a relatively large four-way directional control valve because rod end oil and pump flow combine at the pump outlet and pass through the four-way valve. Note that valve *B* directs fluid back to tank without passing through the four-way directional control valve. This feature can be incorporated in the circuit of Fig. 3.29(a) and (b). The signal functions in the circuit of Figs. 3.29(b) and 3.30 are similar.

Combination Regenerative Valve with Pilot Check

The valve shown in Fig. 3.31 (a) and (b) provides a full regenerative circuit and, in addition, a pilot check valve to lock oil in the blind end of the

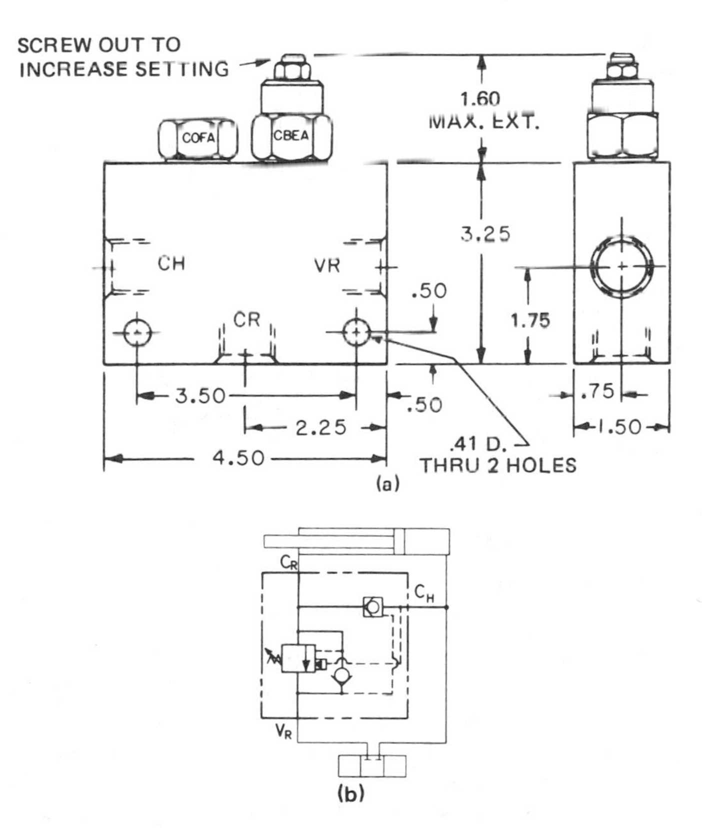

Figure 3.29 (a) Pressure sensitive regenerative valve for rapid cylinder advance. (b) Symbol for regenerative valve and circuit. (Courtesy of Sun Hydraulics Corporation, Sarasota, Florida.)

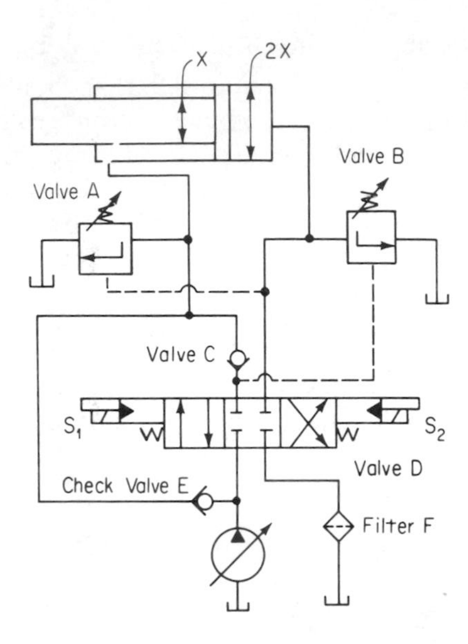

Figure 3.30 Regenerative circuit with separate components.

cylinder. This positively locks the rod in position against any kickback from the load.

The sum of all the flow—pump plus regenerated oil—passes through these valves. Typical regeneration occurs between 500 psi minimum and 1200 psi maximum (adjustable) with a typical maximum pressure of 3000 psi.

Note the manual release facility to unseat the pilot check in an emergency situation.

3.4. HYDRAULIC MOTOR CONTROLS

3.4.1. Hydrostatic Transmission Valves

For Use with Bidirectional Hydrostatic Transmissions

The pump and motor employed in a hydrostatic transmission must have protection from overpressure conditions in each direction of rotation and a source of fluid at relatively low pressure to replace fluid used to lubricate and cool components. An auxiliary pump moves and pressurizes the oil at a relatively low pressure to accomplish these functions. It can also be a source of pressurized fluid for control purposes. Note the circuit of Fig. 3.32(a) and (b). Two major relief valves are included to direct fluid from one line to the second. The charge pump (auxiliary pump to maintain pressure in the supply portion of the circuit) is included in the main pump assembly.

The check valves from the charge pump into each major line are also integrated into this main pump assembly. A relief valve to control maximum charge pump pressure is also included in the main pump assembly.

A suitable directional valve and secondary relief is included in the transmission valve to establish maximum pressure in the make-up circuit bleeding oil which can be used for filtering and cooling the fluid.

For safety reasons a mechanical brake is recommended for positively locking any live loads being handled by the hydrostatic transmission device.

Charge Pump Relief with Shuttle

The valve assembly of Fig. 3.33(a) and (b) is designed for use in closed circuit transmissions which do not have an external charge pump relief. It allows hot oil to be bled from the circuit for cooling and filtration.

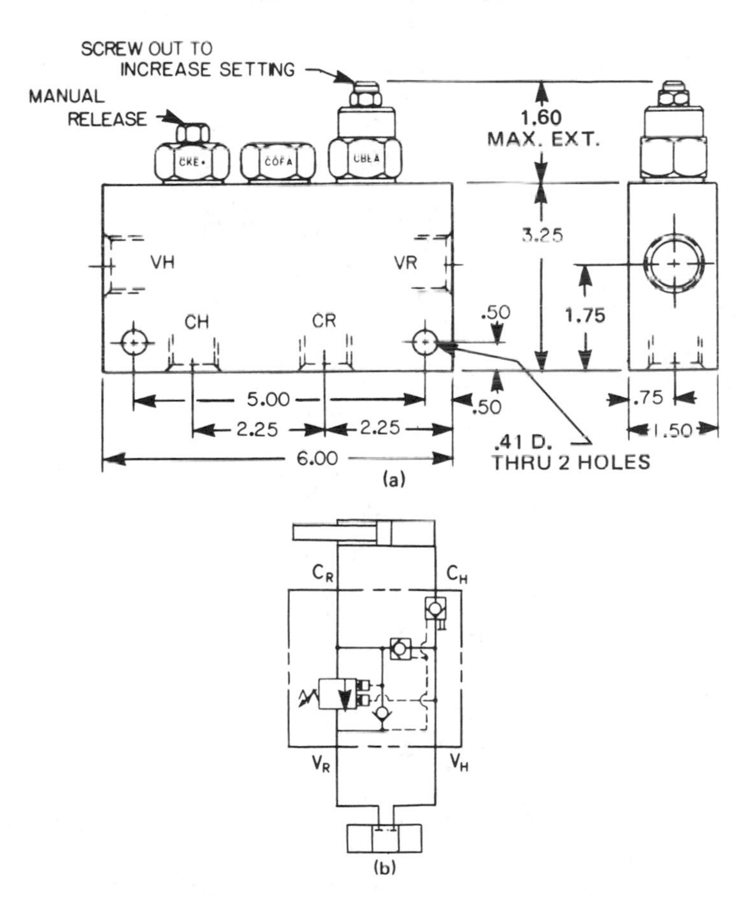

Figure 3.31 (a) Regenerative circuit with pilot check valve. (b) Symbol for regenerative valve with pilot checks. (Courtesy of Sun Hydraulics Corporation, Sarasota, Florida.)

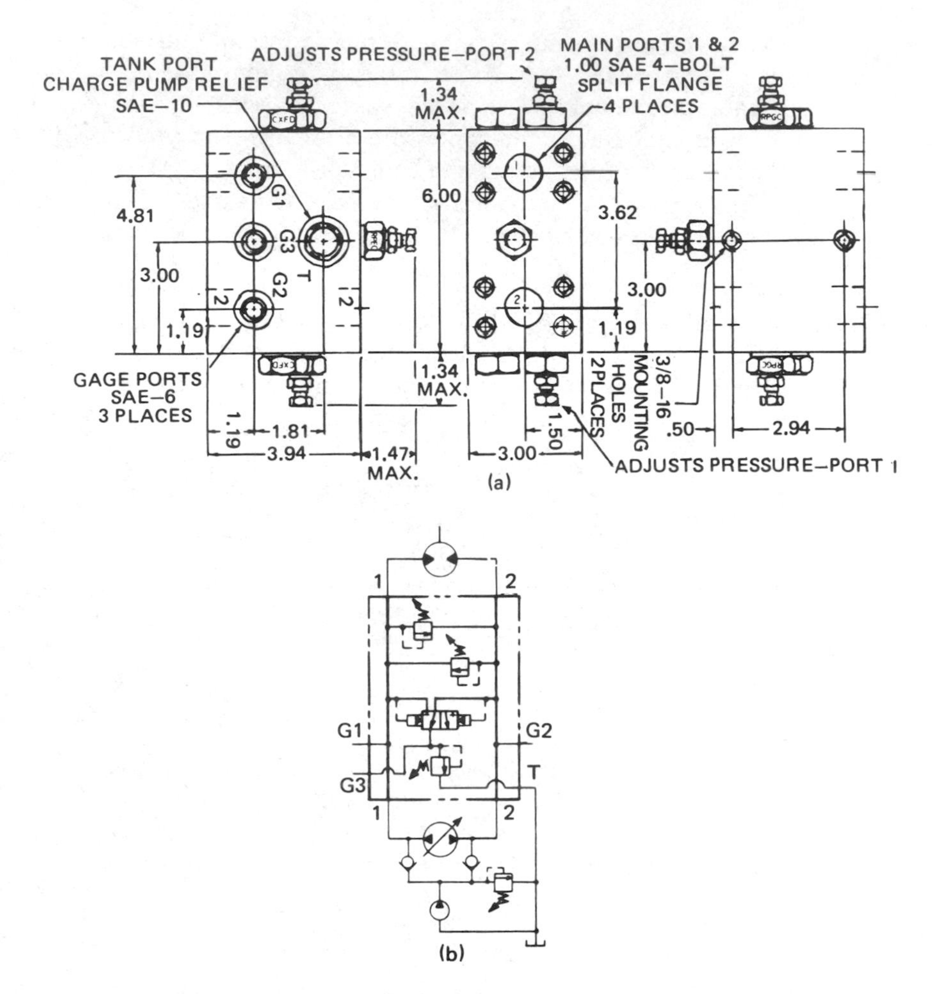

Figure 3.32 (a) Hydrostatic transmission valve. (b) Symbol for hydrostatic transmission valve. (Courtesy of Sun Hydraulics Corporation, Sarasota, Florida.)

Customized Crossover Relief Valves

Crossover relief valves can be housed in a body which can be manifolded to specific motor port connections to reduce plumbing and provide relief at the most effective area in the circuit. The valve of Figure 3.34(a) and (b) also includes a check assembly to allow entry of oil to prevent cavitation in that portion of the circuit.

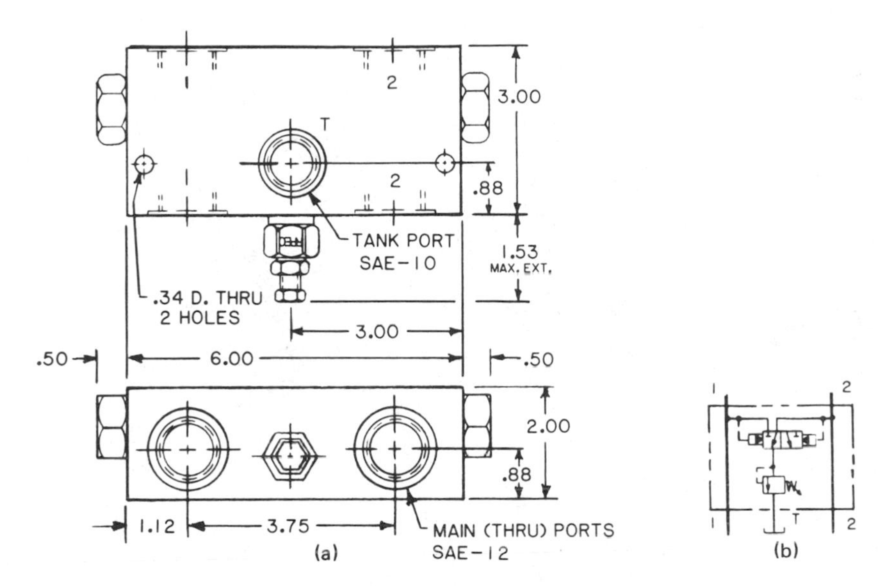

Figure 3.33 (a) Charge pump relief with shuttle. (b) Symbol. (Courtesy of Sun Hydraulics Corporation, Sarasota, Florida.)

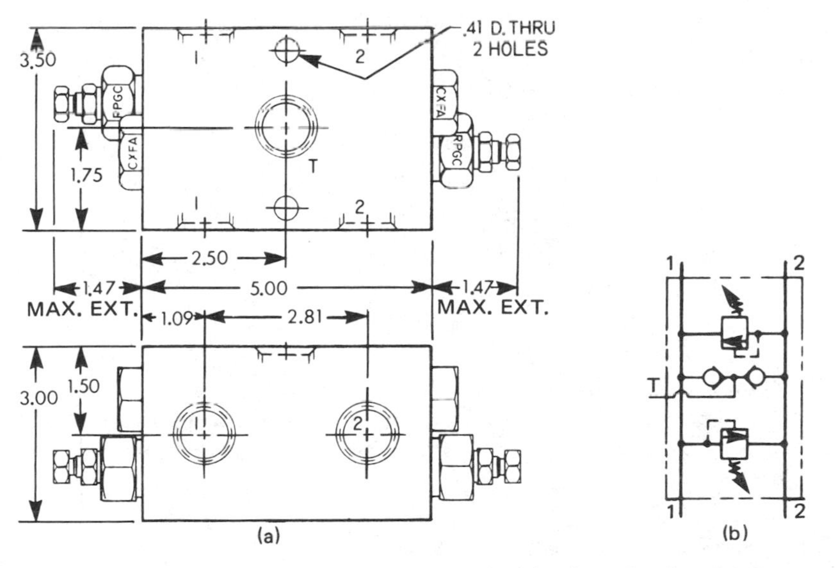

Figure 3.34 (a) Cross-port reliefs with anticavitation checks. (b) Symbol. (Courtesy of Sun Hydraulics Corporation, Sarasota, Florida.)

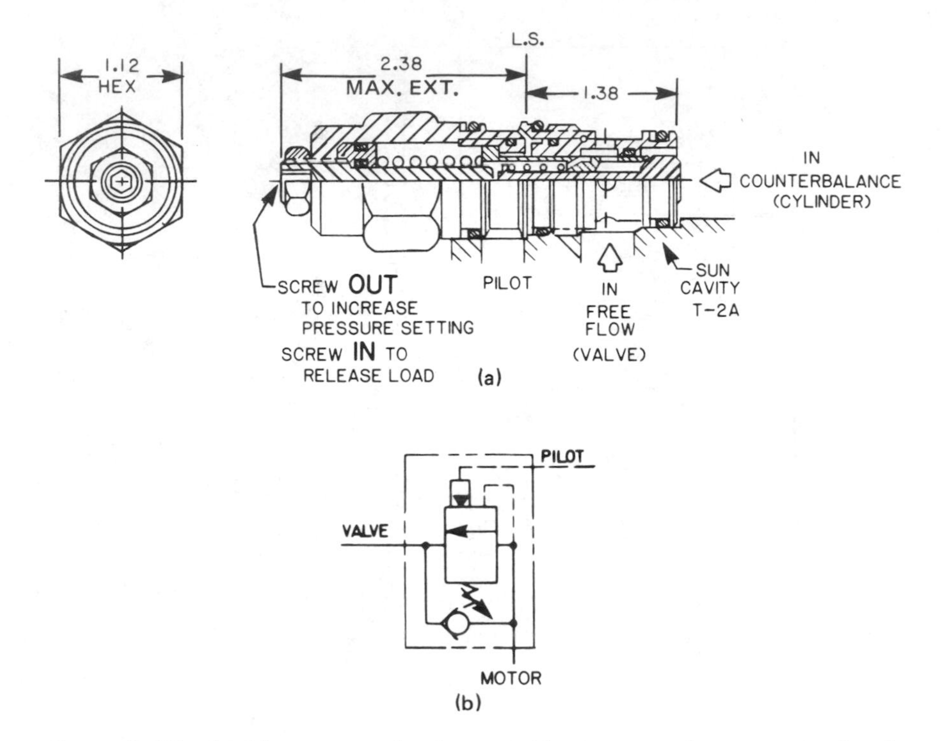

Figure 3.35 (a) Motor control valve cartridge to control overrunning loads. (b) Symbol for motor control valve cartridge. (Courtesy of Sun Hydraulics Corporation, Sarasota, Florida.)

3.4.2. Motor Controls to Control Overrunning Loads

Counterbalance cartridges for motor circuits will provide deceleration control, counterbalancing, load locking, and anticavitation [Fig. 3.35(a) and (b)]. Characteristically, a high pilot ratio (10:1) is used so that the valves operate from a relatively low pilot signal, reducing the power required to overdrive the valves. Motor control valves are available in single and dual function assemblies, to control one or two directions of motion.

Assemblies with an integral shuttle valve, which will provide an external signal to release a spring applied, hydraulically released brake, are also available.

3.5. MISCELLANEOUS VALVES

3.5.1. Air Bleed and Start-Up Valves

Start-up and air bleed valves will reduce power requirements and facilitate pump priming during start-up in a blocked center circuit. It will allow the pump to come up to speed under light load, purging the system of air, before it closes and establishes full system pressure.

The valve consists of a normally-open two-way valve with an orificed flow. The signal upstream of the valve is applied to the piston opposite the bias spring. As solid oil enters the orifice, resistance is encountered, and the spool is urged against the bias spring closing the path to tank and stays closed until the pump flow ceases.

3.5.2. Hot Oil (Hydrostatic with Charge Pump)

Hydrostatic transmissions may have the highest heat concentration in major working lines. The valve of Fig. 3.33(a) diverts charge flow from the low pressure leg of the circuit through the heat exchanger to minimize heat in the major working lines. Filtration can also be accomplished.

3.6. SUMMARY

Pressure control valves can be categorized in four areas. First, normally-closed two-way valves which pass flow at a predetermined pressure level. Second, normally-open two-way valves which close at a predetermined pressure level to provide a stable secondary pressure regardless of the upstream pressure. Third, pressures are translated into a mechanical or electrical signal for purposes of instrumentation or to provide signals for secondary operations, or fourth, the valves are integrated into an assembly with other controls or operators or into the pump structure to minimize piping, reduce cost, save space, or improve circuit operation.

4 Directional Control Valves

4.1. INTRODUCTION

Descriptive titles have been assigned to or accepted from common usage for the hydraulic valves used to control direction of flow in a hydraulic circuit. The use of the word *check* in the context of a machine control brings to mind the cessation of or abrupt interruption of motion. And this, of course, is precisely the purpose of these valves. The term *two-way, three-way, four-way,* etc. provide convenient quick identification of the basic function of the valving mechanism.

Most pressure control valves are essentially two-way valves. They can be normally open when used as a reducing valve or normally closed when used as a safety, relief, sequence, counterbalance, or unloading valve. When two-way valves are used for pressure level control they usually modulate, or constantly change flow rate through the valve, to create an orifice or resistance to flow necessary to establish the desired pressure level in the circuit portion that they control. These two-way valves serve both a directional and pressure level control function.

The choice of a two-way valve for a circuit control function not associated with pressure level control involves several considerations.

As an example, a cylinder with a 2:1 rod in the extending mode with a 10 gpm supply will result in a flow from the rod end at a rate of 5 gpm. In the opposite direction, the 10 gpm supply to the rod end will result in a flow from the blind end of 20 gpm. A 10 gpm rated four-way directional valve can be much less expensive that a 20 gpm valve. A 10 gpm rated two-way valve can be relatively inexpensive. Connecting the two-way valve from the blind end of the cylinder to tank in parallel with the four-way valve and piloting it from the rod end may cost somewhat less than the larger valve and piping needed for 20 gpm. But a second benefit which may be of equal value or greater importance is the lower pressure drop through the less complex two-way valve, greater freedom in positioning lines, and smaller space requirements.

A two-way valve is often used to divert flow through an orifice (Fig. 4.11 below). This function can be controlled by a cam-operated structure as shown in Figure 4.11, a lever-operated mechanism, pilot control by liquid or compressed air, or by electrical means via a solenoid or other motor operator (Fig. 4.14 below).

Two major functions are associated with three-way valves. First is the diversion of pressurized fluid from one common input port to alternate ports as shown in Fig. 4.15 (a). An example would be a tractor equipped with a backhoe and a front-end loader. A single hydraulic system serves both functions selectively. The operator positions the flow directing mechanism within the valve at his option to direct flow to either working mechanism.

The use of a three-way to control a single-acting cylinder is explained later in this chapter.

The use of jet-pipe three-way valve [Fig. 4.15(c)] is explained in Chapter 6 as it is employed in the pilot section of a servo-type directional control valve.

Selection of the four-way directional control valve may be a key to circuit efficiency and long-range dependability. In this chapter we will discuss several available designs of four-way valves and the applications for which they are intended. We will also discuss the choice of four-way valves designed with integrated relief, check, bypass, and similar auxiliary control mechanisms within the four-way assembly.

Valves designed primarily for high production manufacturing systems may be best installed and serviced with interconnecting pipe terminating in a manifold or subplate.

Those valves designed for installation on and control of devices used for mobile, marine, or military applications may be structured with virtually all of the directional control valves either in a single envelope with appropriate cored passages or with individual sections manifolded into a common block with least number of external connecting lines. The lines may be limited to those connecting to the linear or rotary actuators (hydraulic cylinders and hydraulic motors), a supply line, a return to tank line, and possibly a separate drain line for valve structures used for a pilot function.

Usually the choice of directional valves will be predicated by the type of machinery to be actuated and the general area where it is expected to function.

These several types of directional control valves have been designed with characteristics appropriate to the type of machinery to be controlled and the ambient operating conditions. In like manner, basic operational patterns may be provided with completely different mechanical structures. Considerations such as weight, space limitations, ambient operating conditions, ease of maintenance, and initial cost directly affect the design decisions. Significant design flexibility has resulted from the availability of standardized cartridge units which can be integrated into appropriate housings and various manifold and sandwich structures. Multiple monoblock assemblies are available where two or more control devices are integrated into a single block structure (See Fig. 3.4). Individual valves that can be interconnected with suitable pipe, tubes, or hose offer considerable design and fabrication flexibility. There can be some potential cost problems associated with the greater amount of labor involved in the fabrication function.

Valves are also designed to be compatible with the fluid medium deemed best for the planned hydraulic transmission system.

Petroleum based fluids offer signficiant advantages for hydraulic systems. These advantages are related to lubrication, stability, long life, and ready availability. Cost of petroleum-based fluids are relatively high. Potential contamination from leakage and fire hazards are always a concern.

Use of water and/or other chemical fluids may require different seals, protection from oxidation, and possibly lower operating pressures. Obviously, choice of operating fluid is of concern in the selection of all hydraulic components within a hydraulic power transmission circuit. It may be of special concern in the choice of directional valves because of the very precise response associated with the directional control function and their vulnerability to flow forces and power needed to shift the flow directing elements.

4.2 CHECK VALVES

4.2.1. Single Automatic Function

The common household and industrial check valve with a hinged gate-type closure or ball and seat combination finds use in low-pressure areas of hydraulic circuits to allow free flow in one direction with *no* flow in the opposite direction. Often this type of check valve must be installed in a specific manner because gravity is involved in the closure mechanism.

Most hydraulic power transmission systems use check valves incorporating a compression spring to hold the moving member against the seat regardless of installed position.

Type of Fluid Seal

Resilient Seal in Poppet Seat Assembly Elastomers of various types can be mechanically retained in the seat structure of a check valve to provide a deformable gasket mechanism. The cartridge structure of Fig. 4.1(a) illustrates the use of an elastomeric O-ring to provide the sealing function. The cartridge is designed with an undercut for an external resilient O-ring type seal to stop flow at the outer diameter of the cartridge assembly. The cartridge is inserted in a machined bore as shown in Fig. 4.1(b) and retained mechanically. The cartridge can be retained within a subplate assembly as shown in Fig. 4.1(c). A simple pump mechanism can be created by inserting two check cartridges into a cylinder cap as shown in Fig. 4.1(d). When inserted in a manifold block the check can be oriented and retained in any manner appropriate to the cir cuit flow requirements as shown in Fig. 4.1(e) and (f).

Some circuit designs require a check valve in the suction line of the pump. The cartridge structure of Fig. 4.1(g) is inserted into the pump housing to minimize plumbing and reduce cost.

The threaded insert-type cartridge structure of Fig. 4.2(a) is available with a plastic insert-type seat for positive seal in the closed position.

Hard Poppet Seat Assembly The threaded insert-type cartridge of Fig. 4.2(a) is also available with a hardened and ground seat for service where compatibility problems may exist or a metallic seat is preferred.

Biasing Mechanisms

The compression spring of Fig. 4.1(a) is contained within the poppet mechanism as is that of Fig. 4.2(c). The design of the cartridge assembly of Fig. 4.2(a) offers an alternative path to that required in the first two cartridges. The spring is guided by the poppet stem. Thus an alternate pattern is available. Note the choice as shown in the symbol of Fig. 4.2(b) and Fig. 4.2(d).

Cracking Pressure Circuit considerations dictate the required pressure level created by the bias spring. Usually, a low pressure of 5 psi or less may be considered standard in nonreturn flow devices.

Generally the cracking pressure for a check valve is less than 100 psi. Two or more alternate interchangeable spring assemblies to provide desired cracking pressure may be available on a stock basis.

The check valve may also provide a sequence or priority function wherein the cracking pressure may be at a higher value to divert fluid around a saturated filter or provide a minimum pilot pressure source to actuate other valves in the circuit.

Poppet Guide The poppet guide and spring pocket may create an area which can be used to control the actual rate at which the check will open or close. In the basic mode the poppet would be free to move. Restrictive movement would be tailored to specific circuit requirements.

Integral Orifice Type

A drilled hole within the poppet or in the check valve housing assembly can permit free flow in one direction and a restricted flow in the other direction.

(a)

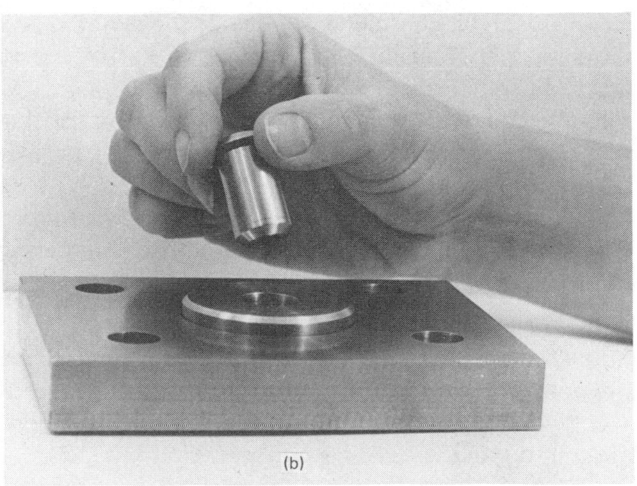
(b)

Figure 4.1 Resilient seat check valve provides minimum flow loss. (b) Cartridge can be installed in a smooth bore. O-ring on OD of valve body provides static seal. (Courtesy of Kepner Products Company, Villa Park, Illinois.)

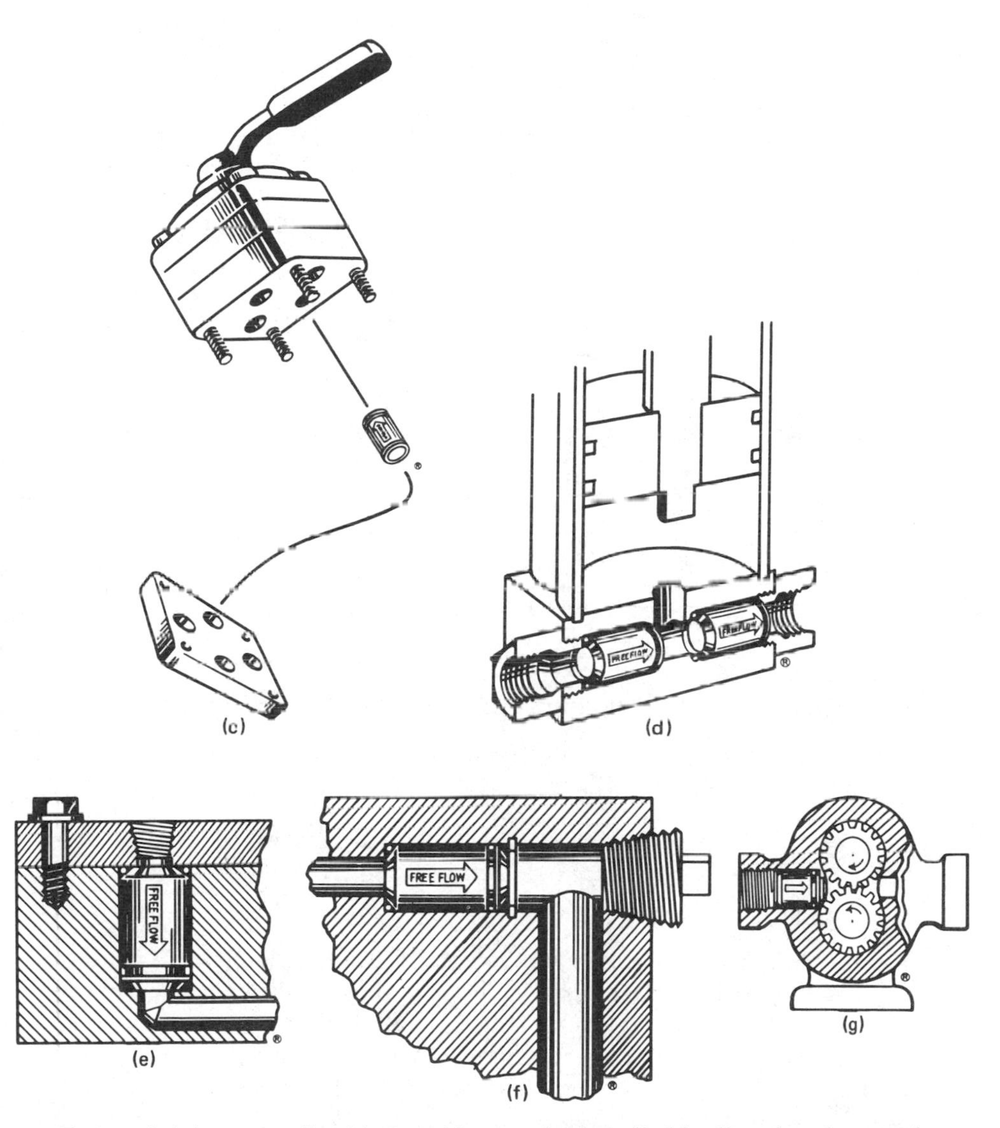

Figure 4.1 *(continued)* (c) Cartridge can be installed in directional control valve assembly. (d) Check valve cartridge in cylinder head can provide controls for a pumping function. (e) Check valve cartridge retained with cover assembly. (f) Check valve cartridge retained with snap ring device. (g) Check valve can be fitted to pump inlet to maintain primc. (Courtesy of Kepner Products Company, Villa Park, Illinois.)

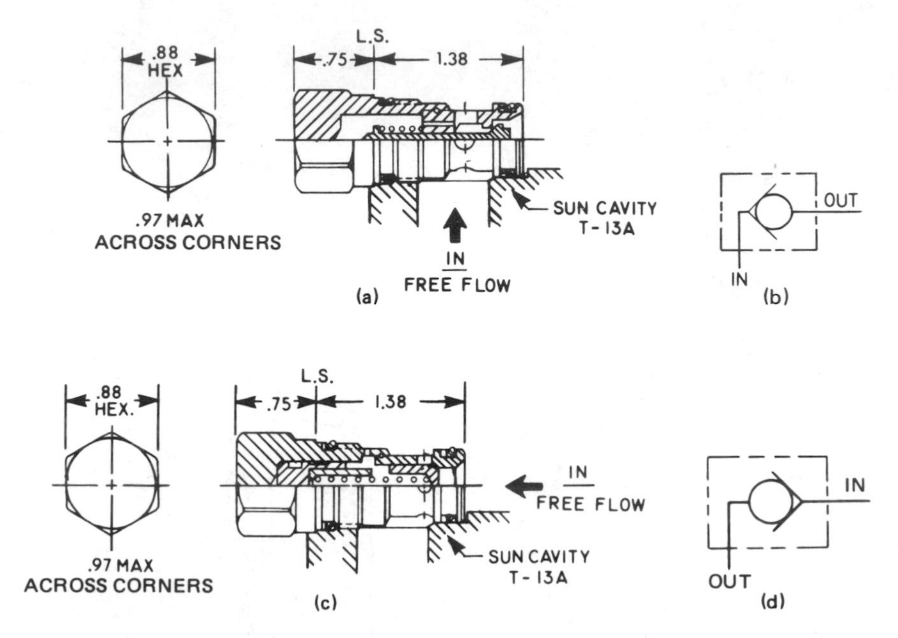

Figure 4.2 (a) Threaded insert cartridge check valve with resilient seat insert. (b) Check valve symbol modified to show flow pattern. (c) Threaded insert cartridge check valve with hardened steel seat. (d) Check valve symbol showing modified flow pattern. (Courtesy of Sun Hydraulics Corporation, Sarasota, Florida.)

Obviously this is no longer a *check* valve in the true sense of the working function but this restrictive device can be structured from a basic check valve.

Envelope Configuration

The simplest check valve for hydraulic power transmission circuits consists of a ball of greater diameter than the hole in which it is seated. The ball can be of any appropriate material ranging from a soft elastomer to hardest steel or ceramics. The envelope can contain many functional devices with the check structure as one element. Common single purpose check valve envelopes may be in line or at right angles as shown in Fig. 4.2(b). Check valves of ball or poppet type can be housed within valve spools, in hydraulic cylinder, or piston assemblies, or any other location where this uniflow device is needed.

Multiple Inlet: Nonreturn Structures

To insure a dependable source of pilot pressure flow from several points within a hydraulic circuit, fluid flow may be directed through a multiple check assembly and a single outlet to the devices to be piloted. Even though the pump may not be running, a source of pilot pressure may still be available because of gravity acting on a load and potential flow from one or more cylinders within the circuit. This check network senses the highest available pressure and blocks all other potential sources.

4.2.2. Pilot Operated to Open

The basic check valve structure provides the essential elements for the pilot operated check shown in Fig. 4.3(a). A concentric smaller size check poppet assembly is added to the basic check structure to provide a reduced diameter passageway to minimize shock as the poppets are forced from the seat by the pilot piston mechanism in the designed pattern.

Pilot Ratios

The area of the pilot piston at a specific fluid pressure level is designed to create a force value needed to urge the poppet structure from the seat, opening a passageway through the valve.

The *seated* area of the poppet, or poppets, determines pressure level needed to unseat the poppets with a specific pressure level urging the poppets to a closed position. The ratio of areas of the poppet piston to the seated poppet areas establishes the pressure ratios needed to create the desired poppet *opening pattern*.

Low Pressure Downstream Resistance to flow reflected to the face of the pilot piston opposite the pilot pressure area decreases the potential force available to unseat the check poppet assembly. This pressure level must be known to accurately assess the pilot pressure needs within the machine cycle.

Prefill Circuits A pilot-operated check valve used in a prefill circuit rarely encounters downstream resistance to flow. The cross section view of Fig. 4.3(b) shows how a pilot-operated check valve can be used to pass major flows to prefill a large ram prior to pressurization as the die is closed.

A pilot-operated check valve used in a prefill mode is often located within the reservoir with a large passageway adjacent to the pilot piston.

Note in the circuit of Fig. 4.3(c) how the solenoid operated four-way directional valve can direct pressurized flow to the ram area as the dies approach the closed position. Entry of pressurized fluid into the ram area stops the gravity and/or vacuum created as the jack ram forces the main ram upward. In the rapid advance mode the ram acts as a large pump.

At the completion of the curing time of the materials within the die, the major three-position four-way valve reverses. Pilot pressure is directed to the prefill valve. At the same time, the two-position solenoid operated four-way valve is released so that pressure can be relaxed at a controlled rate until the pilot pressure can force the poppet open on the prefill-type pilot-operated check valve to allow rapid movement of the piston/ram assembly.

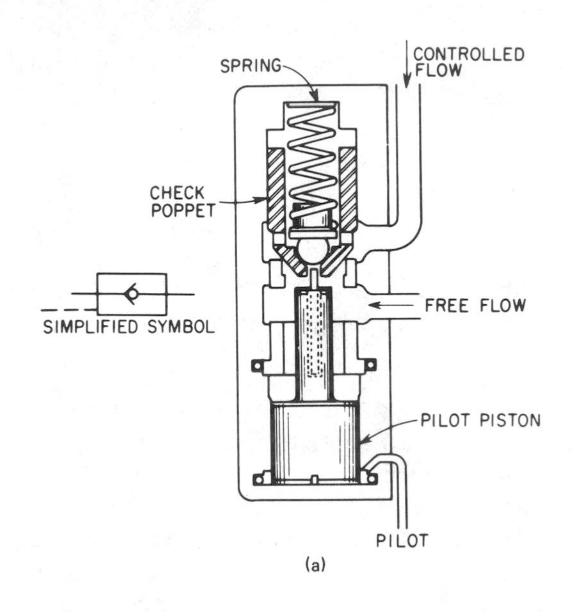

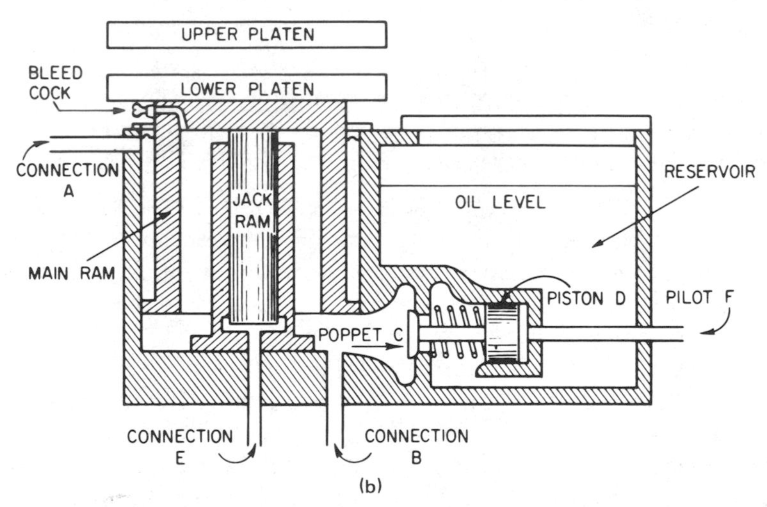

Figure 4.3 (a) Pilot check valve and simplified symbol. (b) Cutaway view of prefill valve. (From Pippenger, John J. and Hicks, Tyler G., *Industrial Hydraulics*, 3rd Edition, Gregg Division, McGraw-Hill Book Company, New York, New York, 1979.)

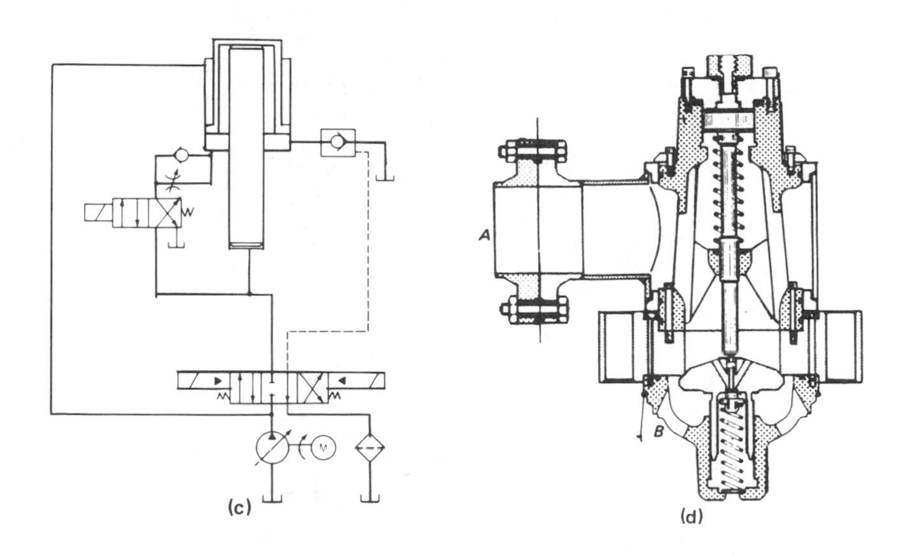

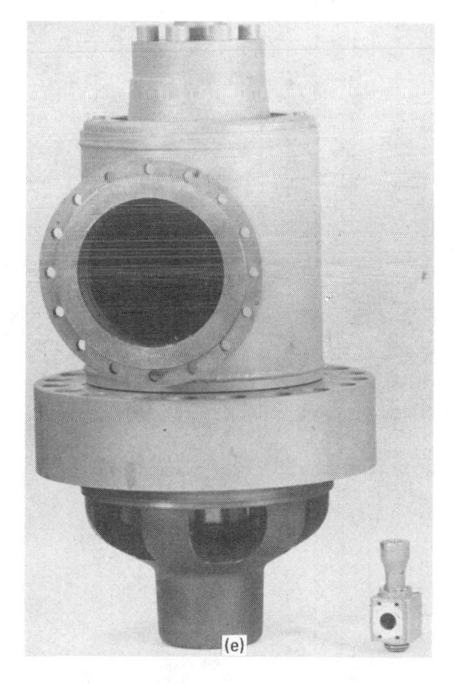

Figure 4.3 *(continued)* (c) Prefill valve circuit. (From Pippenger, John J. and Hicks, Tyler G., *Industrial Hydraulics*, 3rd Edition, Gregg Division, McGraw-Hill Book Company, New York, New York, 1979. (d) Cut away view of prefill valve. (e) External view of prefill valve. (Courtesy of The Rexroth Corporation, Bethlehem, Pennsylvania.)

Pipe connection:	Size 100 (4")	Size 125 (5")	Size 150 (6")	Size 200 (8")	Size 250 (10")	Size 300 (12")	Size 350 (14")
Model A Welding neck flange (DIN 2633)	4"	5"	6"	8"	10"	12"	14"
Models B and K	Tank mounting						
Port X (Models A and B only)	3/4" NPT	3/4" NPT	1" NPT	1 1/4" NPT	1 1/4" NPT	1 1/2" NPT	1 1/2" NPT
Operating pressure range	Port A up to 225 PSI		Port B up to 5000 PSI			Port X up to 5000 PSI	
Cracking pressure	3 PSI						
Hydraulic medium	**Petroleum based hydraulic fluids, phosphate esters**						
Fluid temperature range	-25 to +180°F						
Viscosity range	35 to 1750 SSU						
Mounting position	Any						

(For applications to other specifications please consult us!)

Size	connection flange part no.	gasket part no.	connection flange mounting screws	tightning torque	O-Ring
100 (4")	005 940	002 221	5/8"-11 UNC x 2 1/2" lg	150 (ft-lbs)	2 - 367
125 (5")	007 759	002 222	5/8"-11 UNC x 2 1/2" lg	150 (ft-lbs)	2 - 375
150 (6")	005 941	002 223	3/4" 10 UNC x 2 3/4" lg	260 (ft-lbs)	2 - 379
200 (8")	005 942	002 224	3/4"-10 UNC x 2 3/4" lg	260 (ft-lbs)	2 - 458
250 (10")	002 228	002 225	7/8"-9 UNC x 3" lg	400 (ft-lbs)	2 - 463
300 (12")	002 229	002 226	7/8"-9 UNC x 3 1/4" lg	400 (ft-lbs)	2 - 470
350 (14")	002 230	002 227	7/8"-9 UNC x 3 1/2" lg	400 (ft-lbs)	2 - 473

(f)

Figure 4.3 *(continued)* (f) Technical data for prefill valves from 4 to 14 in. size. (Courtesy of The Rexroth Corporation, Bethlehem, Pennsylvania.)

Calculation of Pressure Required for Pilot Operation

Size	A1 (in^2)	A2 (in^2)	A3 (in^2)	H1 (in)	H2 (in)	F1 (LBS)	F2 (LBS)	$V_{St}X$ (in^3)
100 (4")	15.655	0.394	3.818	0.984	0.748	48 to 66	275 to 503	2.856
125 (5")	23.859	0.589	5.965	1.260	0.906	77 to 104	423 to 714	5.401
150 (6")	33.545	0.760	7.791	1.496	1.063	106 to 157	551 to 928	8.281
200 (8")	57.854	1.491	14.730	1.969	1.299	185 to 280	1044 to 1522	9.135
250 (10")	88.753	2.147	22.187	2.362	1.496	326 to 514	1577 to 2134	33.183
300 (12")	128.583	3.292	33.144	2.953	1.811	480 to 789	2363 to 3317	60.023
350 (14")	179.49	4.985	49.673	3.543	2.087	727 to 1328	3524 to 4956	103.642

Flow related of flow velccity (GPM)

V (ft/s)	Size 100(4")	Size 125(5")	Size 150(6")	Size 200(8")	Size 250(10")	Size 300(12")	Size 350(14")
1.65	62	98	140	251	397	556	767
3.3	124	196	289	500	794	1112	1534
4.95	186	294	420	750	1191	1668	2301
6.6	248	392	560	1004	1592	2224	3068
8.25	310	490	700	1255	1989	2780	3835
9.9	372	588	840	1506	2382	3336	4602
11.55	434	686	980	1757	2779	3892	5369
13.2	496	784	1120	2008	3176	4448	6136

A1 = main poppet area in in^2
A2 = pilot poppet area in in^2
A3 = spool area in in^2
H1 = main poppet stroke in inches
H2 = control spool stroke in inches
pS_tX = pilot pressure at port X in PSI
F1 = force of valve spring in LBS
F2 = force of spool return spring in LBS
V_{St} = pilot volume to open valve in in^3

(g)

Figure 4.3 *(continued)* (g) Calculation of pressure required for pilot operation and flow velocity data. (Courtesy of The Rexroth Corporation, Bethlehem, Pennsylvania.)

Because large flows are expected in prefill valves, the valve is often bolted directly to the large ram [Fig. 4.3(d), (e)] if it is not an integral part as shown in Fig. 4.3(b).

The valve shown in Fig. 4.3(d) bolts to the head of the ram. The connection to the reservoir is by a flange structure. Figure 4.3(e) shows the type of flange structure and the mounting facilities for a typical prefill valve.

Technical data for this typical valve family is shown in Figure 4.3(f). Note that tightening torque values are shown in this data.

Calculations to provide pressure required for pilot operation and flow related to flow velocity is given in Fig. 4.3(g) to insure operation within practical limits as dictated by the physical characteristics of the machine upon which the valve will be used. The allowable velocity will depend on the position of the reservoir, length of lines, and number of bends in the line. Obviously, a higher relative velocity can be tolerated if the valve can be immersed in the reservoir.

Barrier and Drain Significant pressure levels in the cavity adjacent to the inner face of the pilot piston may require the addition of a barrier seal with the area in between being drained as shown in Fig. 4.4. Figure 4.5 shows a typical circuit using a multiple of pilot-operated checks for specific functional purposes.

The transfer-type molding press is fitted with a heavy die. When the

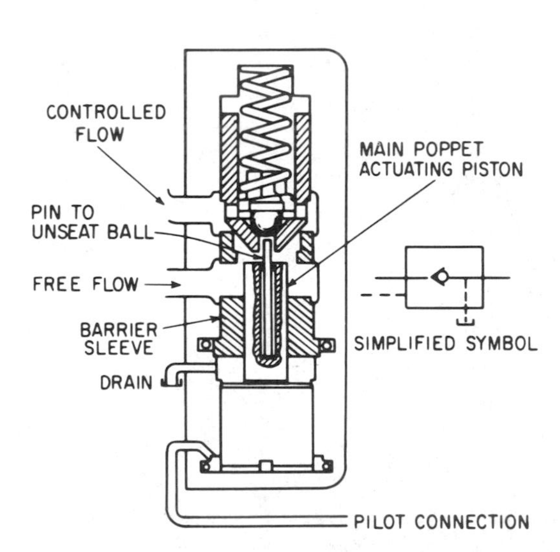

Figure 4.4 Barrier-type pilot check valve and simplified symbol. (From Pippenger, John, J. and Hicks, Tyler G., *Industrial Hydraulics*, 3rd Edition, Gregg Division, McGraw-Hill Book Company, New York, New York, 1979.)

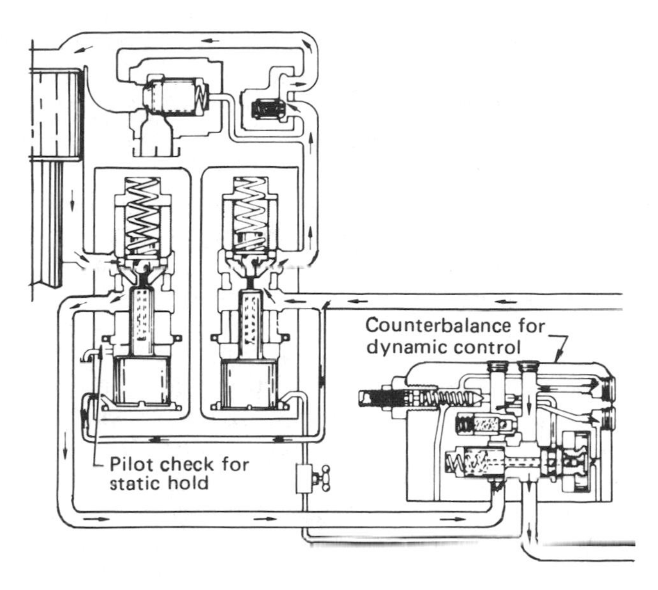

Figure 4.5 Pilot check valve for cylinder lock. (Courtesy of Double A Products Company, Manchester, Michigan.)

press is open and the die is being filled or serviced, the pilot check at the rod end holds the cylinder in position with essentially zero drift. When the directional control is shifted to direct pressurized fluid to the upper end of the cylinder, the quick opening check is piloted closed by the resistance to flow as the liquid passes through the orifice check valve. The counterbalance valve is internally piloted for dynamic speed control to prevent chatter as the lower pilot check is opened.

The resistance of the counterbalance valve necessitates the barrier-type check valve. The die is held closed by the pilot check next to the upper end of the cylinder. As plastic is injected in the die, significant separating forces are exerted on the die.

At the proper time in the cure cycle, the four-way directional valve is reversed, directing pressurized fluid at a controlled rate to the pilot check locking fluid in the head end. Fluid flows freely through the check within the sequence valve and the pilot check adjacent to the piston rod connection.

The pilot check in the line to the blind end of the cylinder relaxes pilot pressure to the normally-closed, piloted-closed, pilot check, which in this installation provides a path to the tank from the blind end of the cylinder with minimum resistance without need to pass through the four-way directional control valve.

Step Poppets Two or more poppets can be concentrically nested in the closure of the pilot check assembly to provide ratios appropriate to decompression needs. A limited quantity of fluid released from the pressurized area will reduce the pressure level to allow operation of the next larger poppet until the major flow is attained.

A cartridge assembly such as that shown in Fig. 4.6(a) can be incorporated in a circuit and identified by the symbol shown in Fig. 4.6(b).

The manual release shown in cutaway Fig. 4.6(c) and symbol of Fig. 4.6(d) can provide a safety release in case of a dead pump, if needed.

Dual Structures

Integration of two cartridges with appropriate cross drilling in an envelope as shown in Fig. 4.7(a) provides control of an actuator in both directions of motion and holds the unit in position until the appropriate pressure signal is received. Figure 4.7(b) shows the identifying symbol.

Figure 4.7(c) shows a dual pilot-check valve designed to be inserted in a sandwich-type structure between a four-way directional control valve and

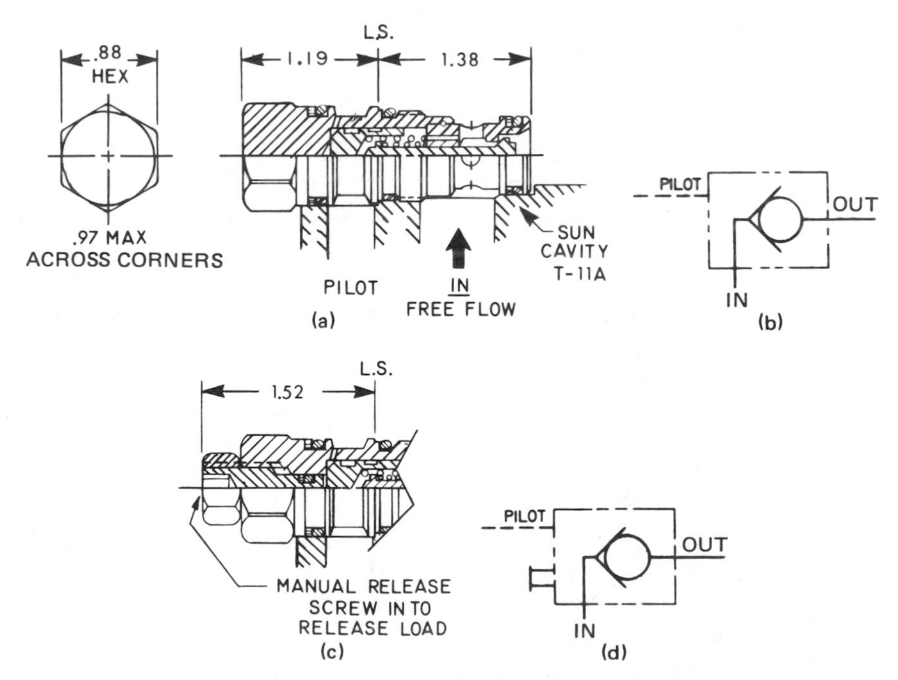

Figure 4.6 (a) Pilot-to-open check valve cartridge (b) Basic symbol for pilot-to-open check valves. (c) Manual release option. (d) Symbol to indicate manual release option. (Courtesy of Sun Hydraulics Corporation, Sarasota, Florida.)

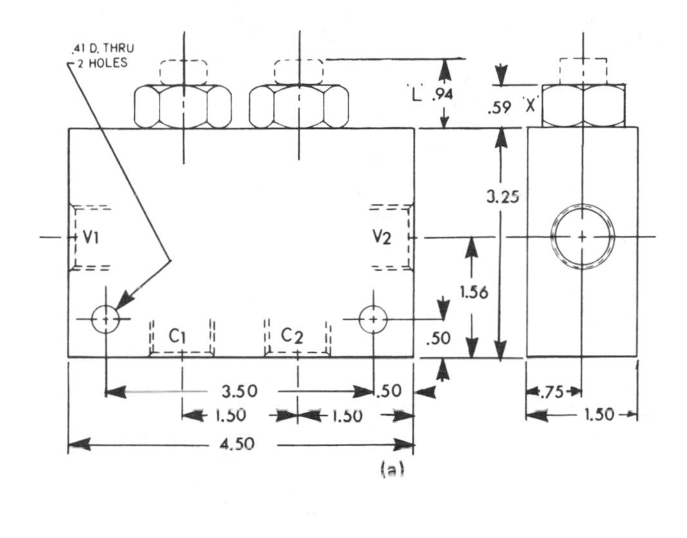

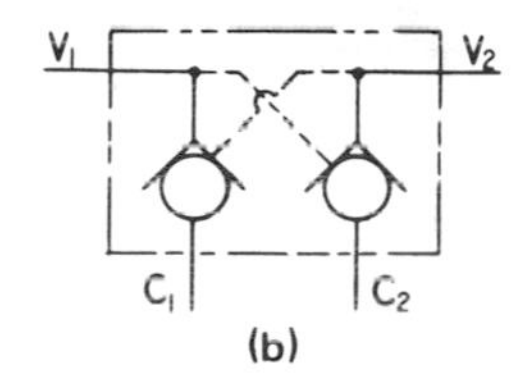

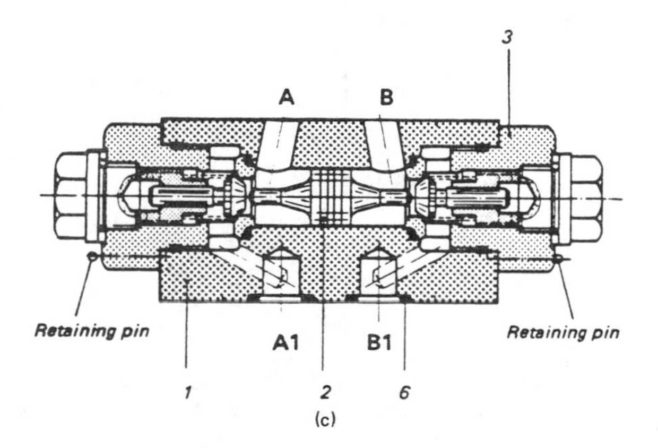

Figure 4.7 (a) Line-mounted dual-pilot check with internal cross pilot. (b) Symbol for dual pilot check. (Courtesy of Sun Hydraulics Corporation, Sarasota, Florida.) (c) Sandwich-type dual pilot check. (Courtesy of The Rexroth Corporation, Bethlehem, Pennsylvania.)

Direction of flow	from A to A1 or B to B1 free flow over check valve: controlled flow from B1 to B or A1 to A
Hydraulic medium	Petroleum Base Fluids
Fluid temperature range	–4 to +160°F
Viscosity range	35 to 1750 SSU
Operating pressure range	up to 4500 PSI
Cracking pressure in free flow direction	7 PSI
Area ratio	$\frac{\text{Poppet}}{\text{Control Spool}} = \frac{1}{2.7}$

(d)

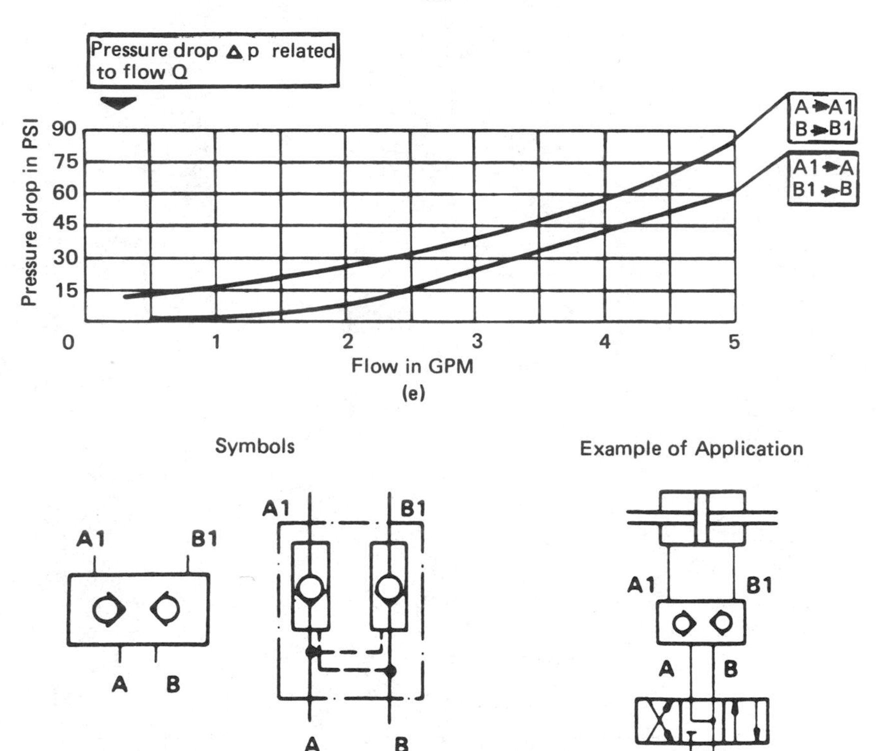

(f)

Figure 4.7 *(continued)* (d) Technical data for dual pilot check valve. (e) Performance curves for dual pilot check valves. (f) Pilot check symbols and example of installation of dual unit. Note the center block of the directional control valve symbol with pressure blocked and both cylinder ports connected to tank. In the center or rest position, the check valve pilots must be relaxed to a low-pressure area or to tank. (Courtesy of The Rexroth Corporation, Bethlehem, Pennsylvania.)

its associated manifold plate to which the fluid lines are connected. Body (1) is equipped with appropriate resilient gasket-type seals. Port A1 and port A2 are locked until pressure is applied at port A or B. Pressure at A can flow freely through the check to port A1 encountering only the resistance of the spring urging the poppet closed as it passes through to the cylinder or motor port. Return fluid is locked until poppet 2 forces the check poppet in housing 3 open, creating a free path from B1 to B and allowing the return fluid an escape route. A similar function is created as pressure is directed to port B.

Technical data for a check assembly of this type is shown in Fig. 4.7(d). Flow characteristics and performance curves are included in Fig. 4.7(e). Figure 4.7(f) shows appropriate symbols and example of application.

The vented unit of Fig. 4.8(a) provides the barrier function to downstream resistance to flow. Figure 4.8(b) shows the identifying symbol.

Envelope Configuration

Pilot checks, like pressure control valves, can be procured in cartridge assemblies, threaded bodies, sandwich structures, and subplate mounted. Most designs are available in both single and dual assemblies.

4.2.3. Piloted to Close

The pilot check adjacent to the blind end of the cylinder of Fig. 4.5 can open in either direction of flow if the pilot line is not pressurized. Pressure needed to open against the face of the poppet is determined by seat area and spring value. To open from the other direction, pressure to the shoulder area must be adequate to compress the bias spring. Pressure to the spring chamber is *additive* to the basic spring.

The cartridge-type *pilot to close* check of Fig. 4.9(a) is made for flow in one direction only. As the symbol [Fig. 4.9(b)] shows, an applied pressure

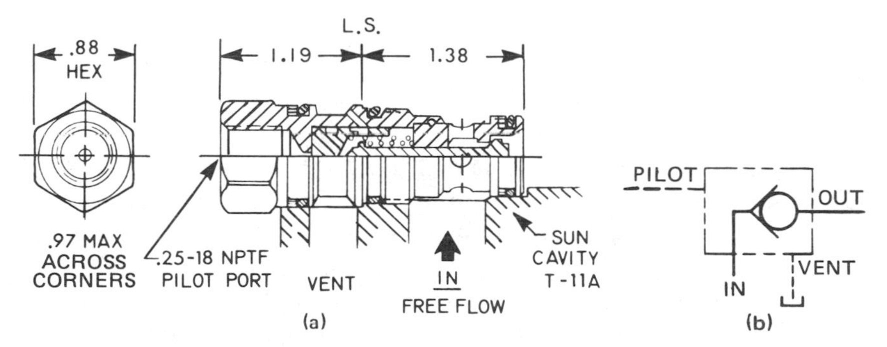

Figure 4.8 (a) Pilot check cartridge with vent function. (b) Symbol showing vent. (Courtesy of Sun Hydraulics Corporation, Sarasota, Florida.)

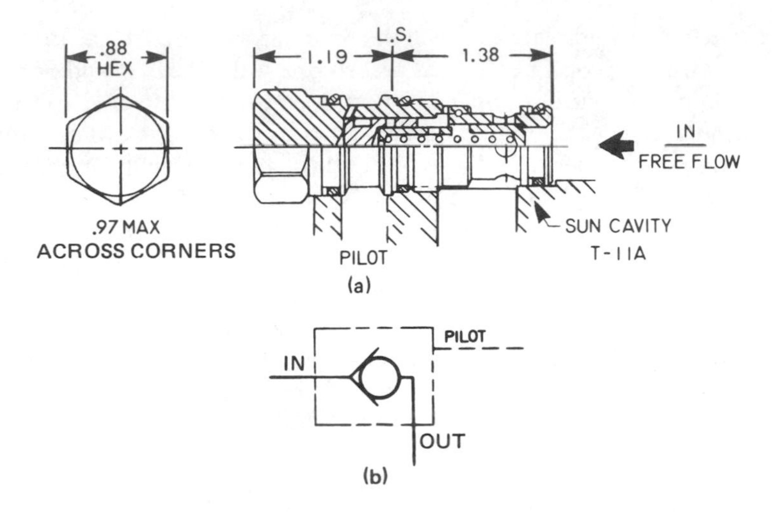

Figure 4.9 (a) Pilot-to-close check valve cartridge. (b) Symbol to show pilot-to-close. (Courtesy of Sun Hydraulics Corporation, Sarasota, Florida.)

will hold the valve closed if the pressure ratios are appropriate to the pilot ratio. A typical ratio is 2:1 wherein 1000 psi will keep the valve closed up to 2000 psi.

4.3. SHUTTLE VALVES

In certain types of machinery, the control must be from more than one point of origin in order to meet circuit requirements.

The three-port shuttle valve in Fig. 4.10(a) can be used to provide a path for fluid from two alternate sources. The shuttle piston will stay in position, blocking flow path to port 1, as long as pressures in port 2 and the outport are greater than the pressure within the entry to port 1. When the pressure at port 1 is greater than that at port 2 and the outlet, it will urge the piston against the stop pin at port 2. This valve will permit flow in both directions when the piston is against the stop pin. The signal to shift the shuttle piston must come from either port 1 or port 2. The signal cannot come from the outport. A definite pressure differential should be available between ports 1 and 2, so that the shuttle piston does not lodge halfway and block the outport in certain circuit applications.

4.3.1. Poppet Type

A ball-and-seat arrangement as in Fig. 4.10(b,c) may be used to provide a leak-tight seal. A fluid flow to all three ports can be expected as the ball passes center in the crossover function.

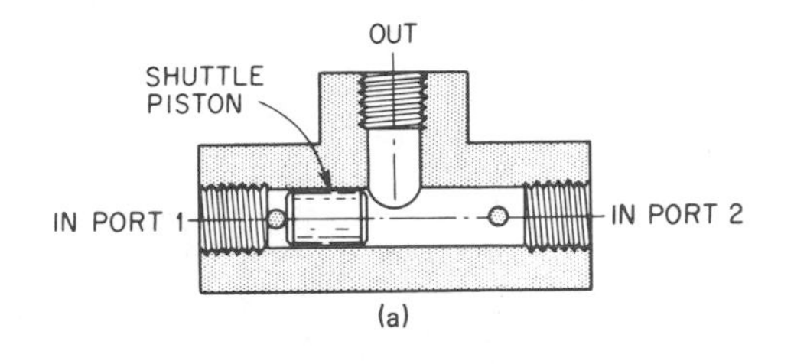

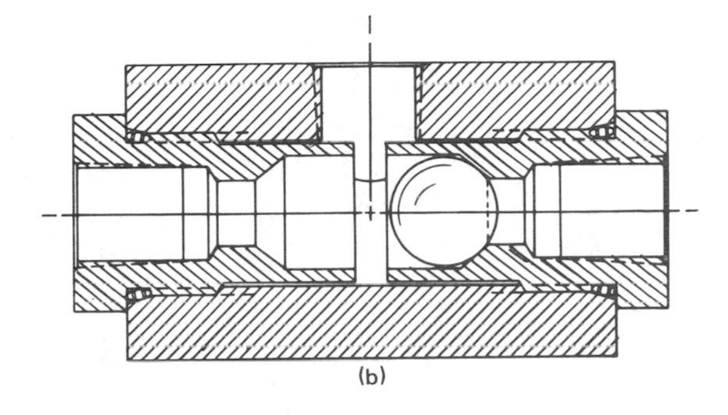

Figure 4.10 (a) Piston-type shuttle valve. (From Pippenger, John J. and Hicks, Tyler, G., *Industrial Hydraulics*, 3rd Edition, Gregg Division, McGraw-Hill Book Company, New York, New York, 1979.) (b) Ball-type shuttle valve. (From Pippenger, John J. and Hicks, Tyler G., *Industrial Hydraulics*, 3rd Edition, Gregg Division, McGraw-Hill Book Company, New York, New York, 1979.) (c) Ball–type shuttle valve with resilient seat. (Courtesy of Kepner Products Company, Villa Park, Illinois.)

4.3.2. Spool Type

The shuttle piston of Fig. 4.10(a) is not completely dead-tight in the operating position. It does not, however, connect all three ports in the crossover position. The ball-type unit is widely used because of the economical construction and the tight seal obtained as the ball abuts the seat. The small loss on crossover is usually of limited importance.

4.3.3. Poppet Type with Resilient Seat

The resilient seat provided in the shuttle valve of Fig. 4.10(c) eliminates possibility of signal loss because of annular wear on the poppet ball. Obviously, the resilient seal material must be compatible with the fluid medium. The resilient seat minimizes the hammer action on rapid shifting devices.

4.4. TWO-WAY VALVES

4.4.1 Shut-off Type

With Metering Capabilities

Needle Valves The term *needle* valve [Fig. 4.12(a)] describes a shut-off valve with relatively long adjustment travel prior to closing the passageway from one port to the other. The long travel (Fig. 4.11) provides a degree of flow sensitivity with less change in flow per rotation of adjustment than what would be encountered with other shut-off valves.

Globe Valves Function of a *globe* valve [Fig. 4.12(b)] is much like the needle valve [Fig. 4.12(a)] except that the rate of flow change is much greater for

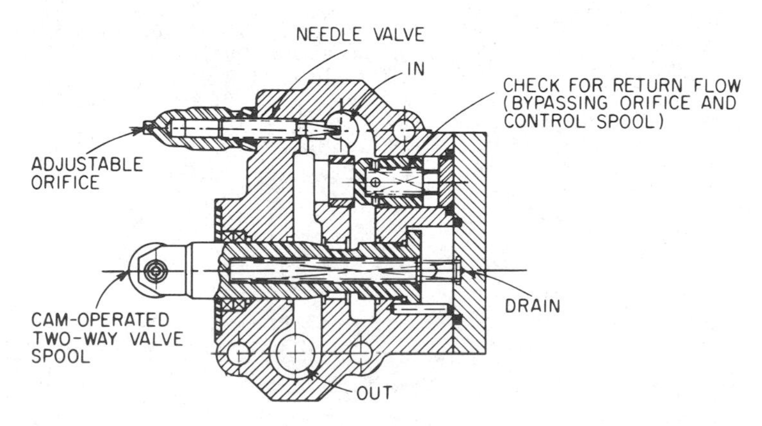

Figure 4.11 Cam-operated two-way valve with check and needle valve. (Courtesy of Double A Products Company, Manchester, Michigan.)

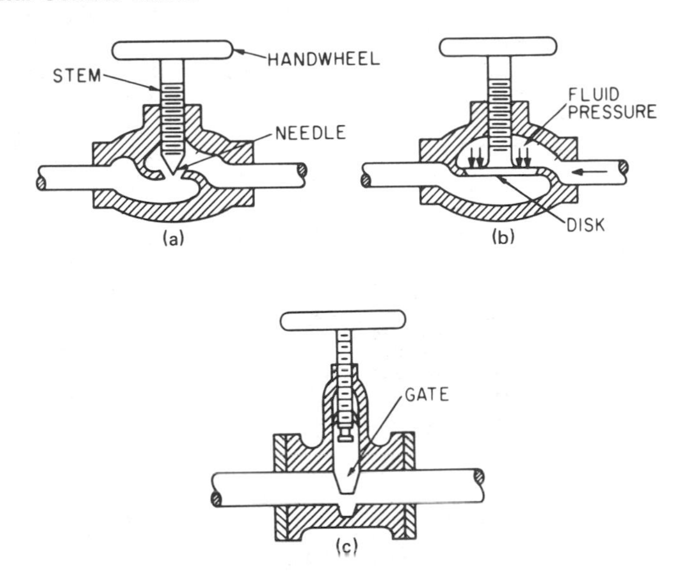

Figure 4.12 (a) Needle Valve (b) Globe Valve (c) Gate Valve (From Pippenger, John J. and Hicks, Tyler G. *Industrial Hydraulic*, 3rd Edition, Gregg Division, McGraw-Hill Book Company, New York, New York, 1979.)

each turn of the operating stem. Usually a much larger flow path is provided than with the needle valve structure in equal housing sizes.

On-off Type

Ball and Plug Valves Complete shut-off to wide open is accomplished by 90° rotation of a ball or plug-type valve (Fig. 4.13). Little throttling is anticipated. The function is basically to open and close a port with minimum motion. Cross section is shown in Fig. 4.13(a), external view of a four-way ball-type valve in Fig. 4.13(b), and symbol in Fig. 4.13(c).

Gate and Butterfly Valves Flow through a valve with minimum pressure loss is important for some low-pressure lines. Gate and butterfly valves provide minimum resistance to flow as do some ball valves. The gate valve [Fig. 4.12(c)] retracts the closure into the valve housing by means of a screw-type actuator. Several turns of the handle are required for full open to full closed position.

A butterfly valve, like a ball valve, opens fully in 90° of rotation. The closure remains in the fluid path as the butterfly valve reaches the full open position. Butterfly valves are rarely used in high-pressure circuits.

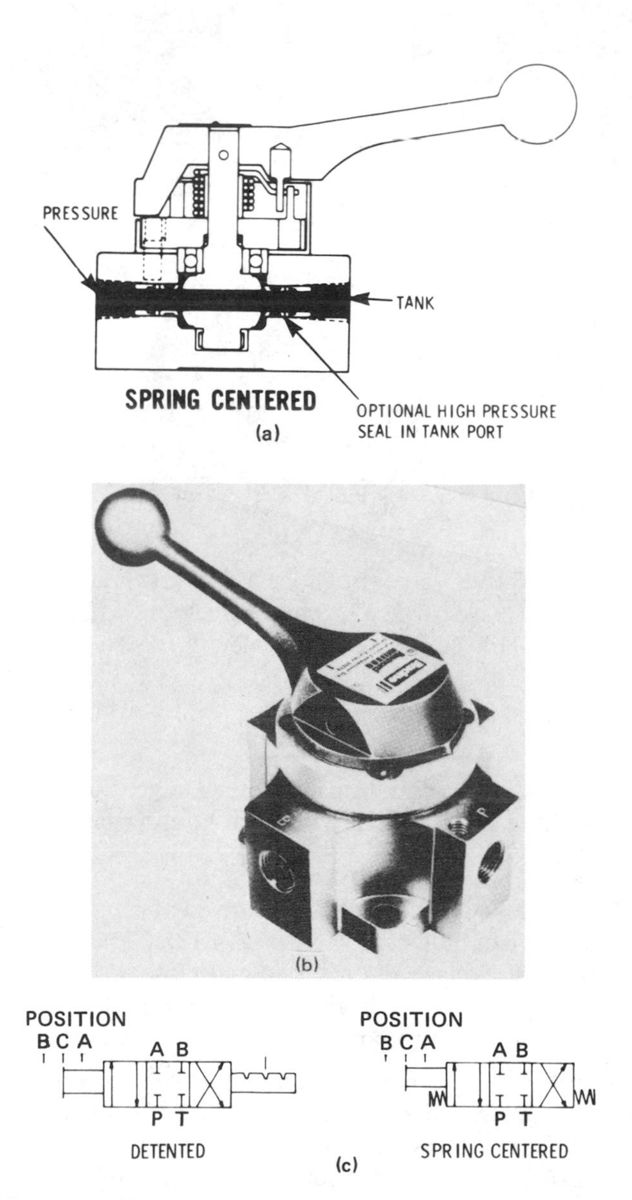

Figure 4.13 (a) Ball-type directional valve. (b) Four-way ball valve. (c) Symbol for four-way ball valve. (Courtesy of Rexnord Fluid Power Division, Racine, Wisconsin.*)

*Now a division of Dana Corp.

4.4.2. Poppet Type

Small flow capacity solenoid operated two-way valves are used to pilot larger valve mechanisms as well as to control flow in hydraulic circuits using low flows. As an example, the solenoid used to vent the relief valve of Fig. 3.10(a) provides a normally-open passageway through the valve [Fig. 3.10(b)] when the solenoid is deenergized. Current to the solenoid moves the flow directing mechanism to block the pilot flow.

At the closing of the vented passageway, the pressure rises to meet the resistance in the circuit. If the resistance in the circuit in high enough, the relief valve will direct excess fluid to tank to maintain the desired preset maximum pressure level. In this type of installation, the two-way valve could be poppet type or spool type with equal performance characteristics. The choice may be based on economic or space considerations.

The solenoid two-way valve used in the normally-closed configuration of the valve shown in Fig. 3.11(a) will create a path to the tank for the relief pilot fluid only if the solenoid is energized.

Frequently it is desirable to relax some portion of a circuit or the entire circuit in a normal machine cycle by energizing a solenoid.

Note the flow path shown in Fig. 3.10(c). This flow, inverse to that of Fig. 3.10(b), offers the designer a choice of controls appropriate to the circuit requirements.

4.4.3. Spool Type

Figure 4.14 illustrates a typical solenoid operated two-way valve designed with a spool-type closure and flow-control element which can be structured for normally-open or normally-closed service.

Relatively large flows can be accommodated by spool-type valves with given actuation forces. A manually controlable override can be provided with this type of valve which may be mandatory in certain installation criteria.

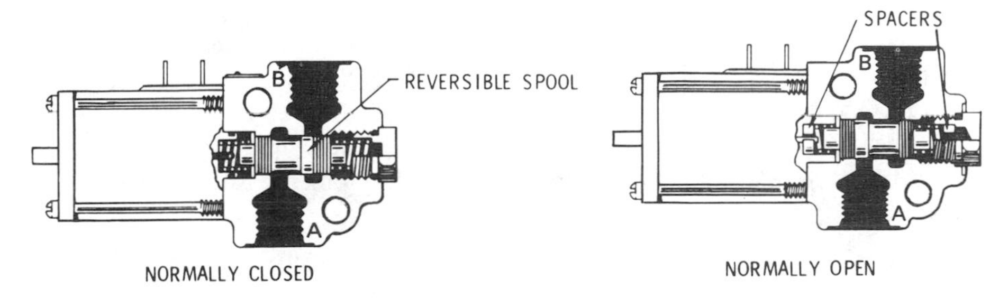

Figure 4.14 Direct solenoid operated two-way valve. (Courtesy of Rexnord Fluid Power Division, Racine, Wisconsin.)

4.5. THREE-WAY VALVES

4.5.1. Plug and Ball Type

The mechanical construction of plug and ball-type three-way valves is similar to that of the two-way version. Obviously, the third port is added. The objective, when using a three-way valve, is to direct flow from one key port selectively to one or the other of the remaining two ports.

Usually a plug or ball-type valve will have a handle pattern moving the flow directing element 180° to create the desired flow paths through the valve.

4.5.2. Spool Type

Diversion Valve Function

One pump can supply two or more circuits under certain circumstances. As an example, a tractor equipped with a front end loader may also have a backhoe at the opposite end of the machine. One pump can supply each circuit selectively as mentioned in the introduction to this chapter.

A valve, similar to that shown in Fig. 4.15(a), is capable of directing the pump delivery to circuit one or to circuit two at the option of the operator.

Single-acting Cylinder Control

For control of a single-acting cylinder, it is only necessary to understand that the input port, by the action of the valve, can be connected to the input of the single-acting cylinder to cause a predetermined movement pattern. The cylinder may be returned by gravity, spring force, or some other means as the valve is shifted in the other position, blocking pressure and connecting the cylinder port to tank. Thus, when used as a single-acting cylinder control, the spool of Fig. 4.15(a) is shifted in one direction to permit the single-acting cylinder rod to travel outward. By shifting the valve in the other direction, the cylinder port will be directed to tank and pressure will be blocked so that the rod can retract in response to the associated energy source.

4.5.3. Poppet Type

Poppet-type valves can minimize leakage in the various shifted positions or in the rest position. Note the structure of Fig. 4.15(b). Pressure holds the poppet closed until poppet A is depressed, permitting flow to the cylinder port. By opening poppet A and B together, the pump and cylinder port can both be relieved to tank. Opening poppet B alone will block pressure and connect the cylinder port to tank.

4.5.4. Jet-pipe Type

The jet-pipe-type three-way valve is generally used for control flow such as the first stage in a servovalve. The pivot point of the jet pipe is supplied with pressurized fluid. The jet pipe moves with modest input energy to direct fluid

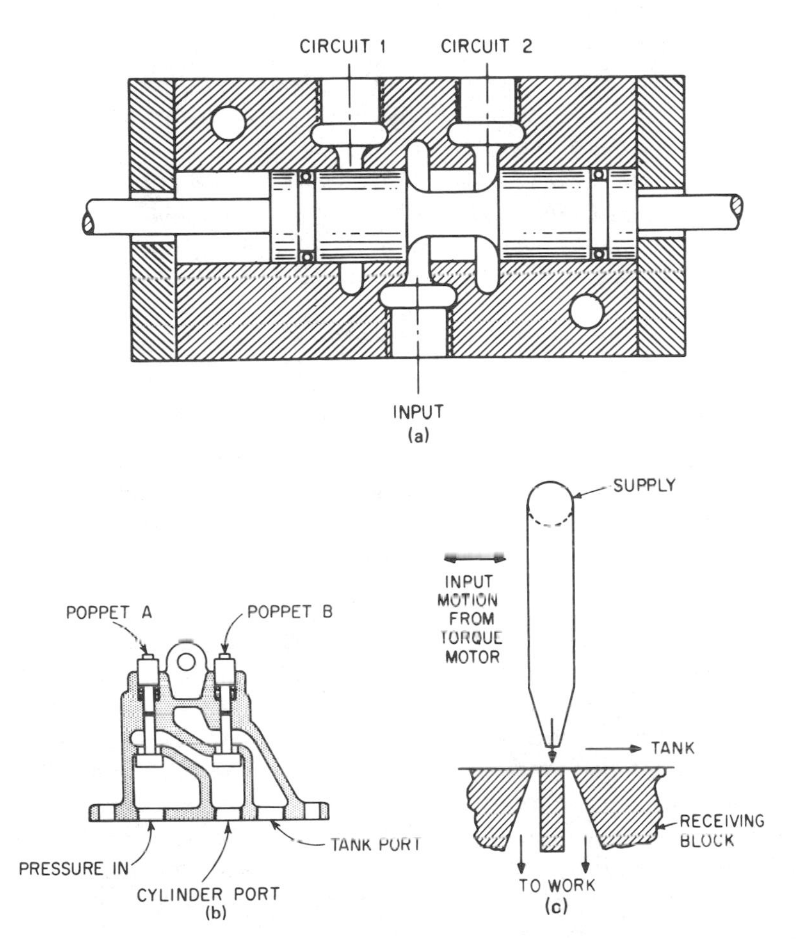

Figure 4.15 (a) Spool-type diversion valve. (b) Three-port, poppet-type, three-way valve. (c) Jet-type three-way valve. (From Pippenger, John J. and Hicks, Tyler G., *Industrial Hydraulics*, 3rd Edition, Gregg Division, McGraw-Hill Book Company, New York, New York, 1979.)

to the desired port as shown in Fig. 4.15(c). The use in a servo-type valve is explained in Chapter 6. The pipe can be moved by a direct current solenoid or force motor.

4.6. FOUR-WAY VALVES

4.6.1. Flow Patterns

The description of the valve symbols in Chapter 2 explained many of the potential flow patterns that can be provided in a four-way control valve.

Finite Positions

A minimum of two finite positions are incorporated in a four-way directional control vavle. These two basic finite positions provide the needed flow paths to reciprocate a linear cylinder or reverse rotation of a fluid motor.

Actual conditions as the spool shifts from one finite position to the other are shown below in Fig. 4.31(b) for a typical valve designed for use with nominal ¾ in. pipe size. This chart shows flow characteristics of the valve as it shifts from one extreme position to the other for both two-position and three-position valves. This information can be essential in choosing the most effective valve for a specific circuit requirement.

Neutral Characteristics

Figure 2.29 shows some typical flow patterns or blocked ports that can be provided in a third neutral or midposition of the four-way flow directing valve.

The chart of Fig. 4.31(a) adds many flow characteristics in addition to the common configurations of Fig. 2.29. The identifying letters are a part of the manufacturers model number code.

Four-position Four-Way Valves

A bulldozer requires a directional control for the blade that includes the expected raise and lower with power. In addition, it may be desirable to let the blade float freely when performing certain operations. This can be accomplished by a four-way directional valve spool which has a finite position where the cylinder lines are connected to each other and to tank to permit the blade to move in response to external forces. This float function is usually considered the fourth function. The normal third function is that of locking the cylinder ports at a chosen position of the blade to accomplish certain grading functions. This requires at least four separate positions of the flow directing mechanism in the bulldozer-type mobile directional control valve [Fig. 4.27(d) below].

Infinite Positions

The parallel lines adjacent to the symbol as shown in Fig. 2.23(i) indicate that the valve may be provided with an infinite number of positions from one completely shifted position to the other completely shifted position.

Some directional control valves assume flow regulating functions as well as the basic directional activities. The infinite positions permit a selective rate of flow as well as discrete direction of the flow.

4.6.2. Plug and Ball Type

Four-way directional control functions can be accomplished with relatively simple plug and ball-type valves [Fig. 4.13(b)]. Usually this type of valve is used where frequent operation is not anticipated.

4.6.3. Disc Type with Biased Sealing Members

The balanced-disc-and-plate construction may have a fixed amount of unbalance to ensure a satisfactory seal. In Fig. 4.16(a), piston-seal assembly abuts a plate containing passages. When the seat is in alignment with some of these passages, a flow is directed with a minimum of resistance through the valve. This piston seal assembly is initially urged toward the mating plate by a wave spring device as shown in Fig. 4.16(b). This provides a seal at relatively low pressures. As the pressure increases, the area created by the outer diameter of the seat minus the inner diameter of a lip on the seat will be effective in holding the seat against the plate. The area of the lip will determine the amount of force holding the two surfaces together. The sliding members are provided with a fine finish to minimize flow between the sealing surfaces. Seals of various types may be used to prevent loss of fluid between the piston seat and the bore in which it is located.

4.6.4. Poppet Type

Many aircraft-type valves are designed with a multiplicity of poppets. These poppets are mechanically forced from their seats by suitable cam action so that the desired flow is produced. The poppet construction provides a tight seal as opposed to the needed clearance flow with the spool-type construction.

The poppet may not provide the smooth acceleration and deceleration characteristics that are possible with the spool-type construction. Sliding-plate and disc construction with various types of balance factors can provide less clearance flow than spool-type units.

Logic systems is a term used to describe a family of controls that use poppet structures to direct fluid flow in a desired pattern to create a two-way, three-way, or four-way function. The logic systems can use pressure, flow, and directional control actions to provide a desired control system.

Relief, sequence, unloading, and counterbalance valves can use a normally-closed poppet structure controlled by a spring-biased pilot valve. Reducing valves can use a normally-open cartridge structure in like manner as shown in Fig. 4.17(a).

The cartridge structure of Fig. 4.17(b) can be fitted into a suitable housing to respond to signals from various points. Note in Fig. 4.17(c) the area pattern that can be created. Figure 4.17(d) and (e) illustrate alternate area patterns that can be provided. The logic cartridge for a normally-closed pressure control function is illustrated by Fig. 4.17(f).

An adjustment can be included to limit the stroke of the poppet as it opens. Figure 4.18 illustrates how four logic poppets in a suitable structure can be used in a four-way mode to control very large flows with a small pilot valve. Large flows often mean more than 100 gpm.

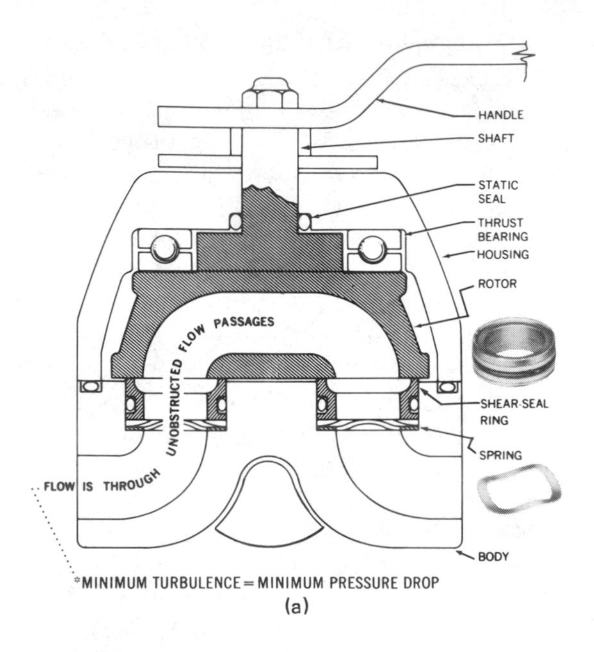

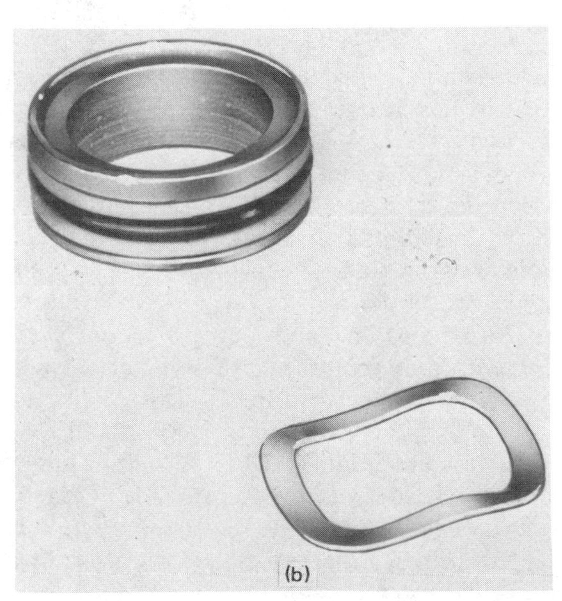

Figure 4.16 (a) Rotary-disc, seal-type, four-way valve. (b) Disc assembly for rotary four-way valve. (Courtesy of Transamerica Delavel, Inc. Barksdale Controls Division, Los Angeles, California.)

The shifted position of the pilot valve of Fig. 4.19 shows how the poppet responds to the pilot flow.

A poppet-type valve assembly can be composed of materials compatible with nonlubricated fluids such as water. A spool-type pilot valve may not function dependably with water. Thus, a poppet-type pilot valve such as that illustrated in Figs. 4.20 and 4.21 may be used.

The poppet-type three-way valve can be used as shown in Fig. 4.22. Figure 4.23 shows an external view of such a valve.

With one solenoid energized as shown in Fig. 4.24, pressure will be directed to A, and B will be directed to tank. To direct P to B and A to T, the solenoid on the left pilot valve is energized as shown in Fig. 4.25. By simultaneously energizing both solenoids, all poppets are free to lift, providing an open center condition as shown in Fig. 4.26. When both solenoids are deenergized, as in Fig. 4.22, there is a fully-closed condition with all poppets held closed.

The logic element structure revitalizes some of the early concepts which led to development of the hydraulic industry as we know it today. In the infancy of hydraulics, water was the medium used to transfer energy. Spool-type directional control valves found universal application only after the introduction of oil hydraulics many years later.

Since economic conditions are again causing us to consider water as our hydraulic medium, many manufacturers are reconsidering the poppet valve as the controlling element. In combining the older poppet theory with modern technology, we find that the logic element provides economical solutions to hydraulic systems incorporating either water or oil at high or low pressures and high or low flows.

4.6.5. Spool Type

Linear Spool and Bore

Spool-type four-way valves are widely used in hydraulic power transmission systems. The versatility and control capabilities (both direction and flow) are tempered by the limited capability to seal a passage completely. In those applications requiring a tight seal, there may be need to use a ball, poppet, or disc-type valve structure. In the design and manufacture of a spool-type valve, there must be some clearance between the spool and associated bore to allow movement. Thermal changes are generally inevitable so that some metal change can be anticipated. The slight clearance in a spool-type valve compensates for thermal changes and in some cases the fluid leakage provides lubrication for ease of movement of the flow-directing spool.

Threaded or Flanged Body – Single Spool Hydraulic circuits can be assembled with individual components for flexibility and potential reuse of the materials.

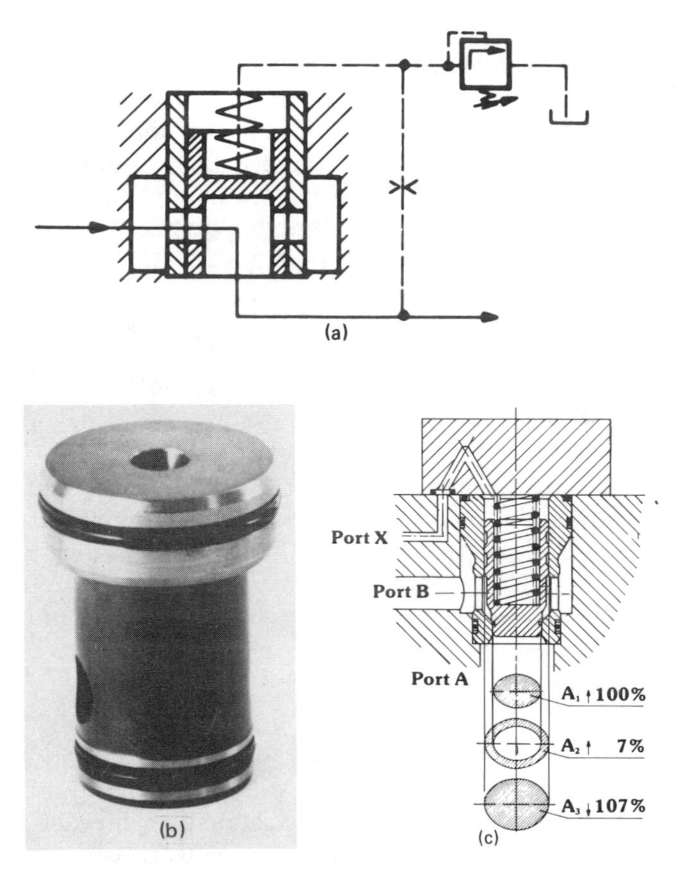

Figure 4.17 (a) Normally-open cartridge assembly. (b) Logic cartridge and cover. (c) Logic cartridge area combinations. (Courtesy of The Rexroth Corporation, Bethlehem, Pennsylvania.)

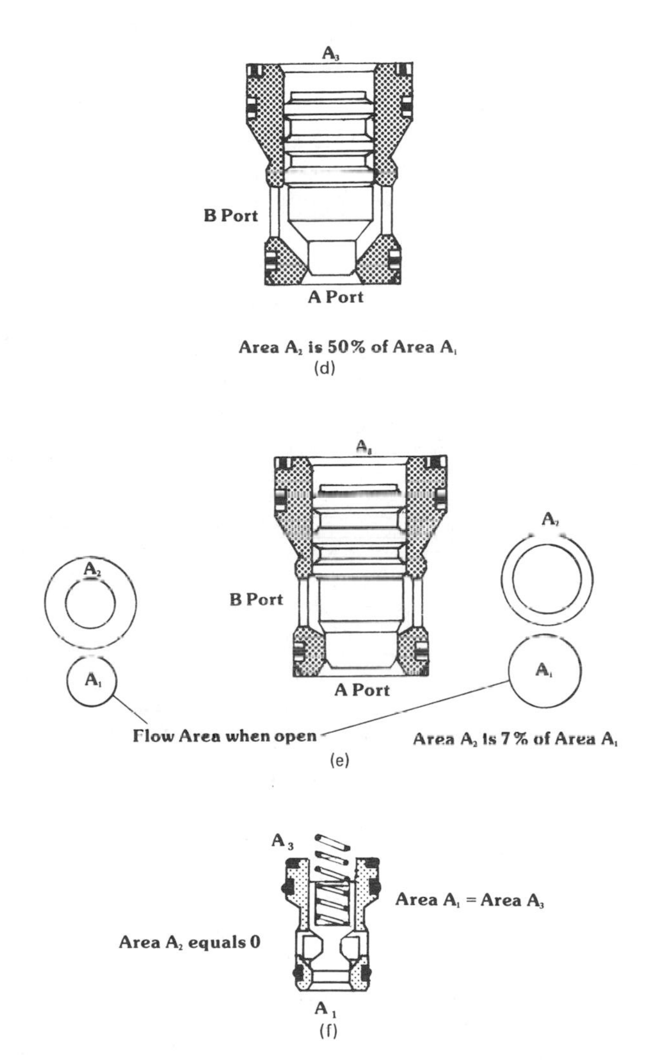

Figure 4.17 *(continued)* (d) Logic cartridge with area A_2 50% of A_1. (e) Logic cartridge with Area A_2 7% of A_1. (f) Logic cartridge for normally-closed pressure control functions. (Courtesy of The Rexroth Corporation, Bethlehem, Pennsylvania.)

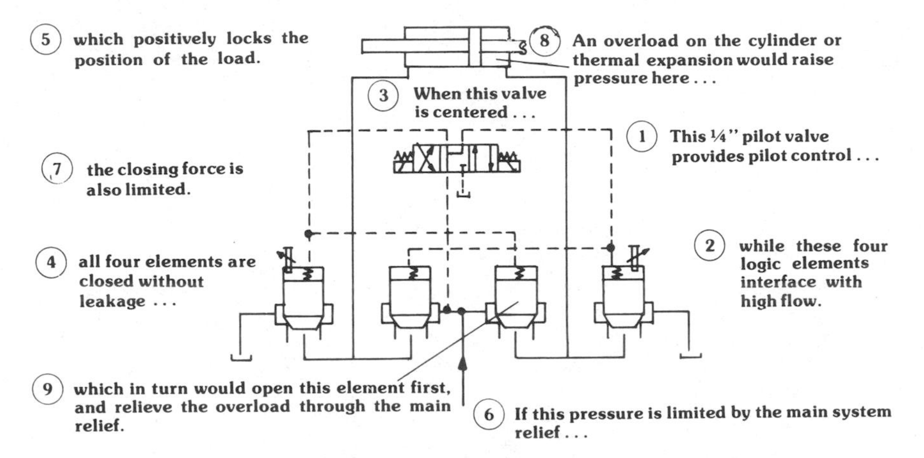

Figure 4.18 Logic circuit, solenoids deenergized. (Courtesy of The Rexroth Corporation, Bethlehem, Pennsylvania.)

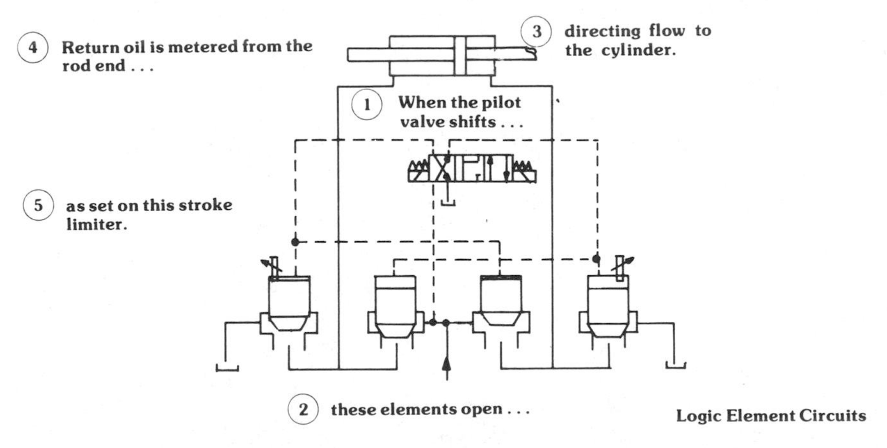

Figure 4.19 Logic circuit, with left solenoid actuated. (Courtesy of The Rexroth Corporation, Bethlehem, Pennslyvania.)

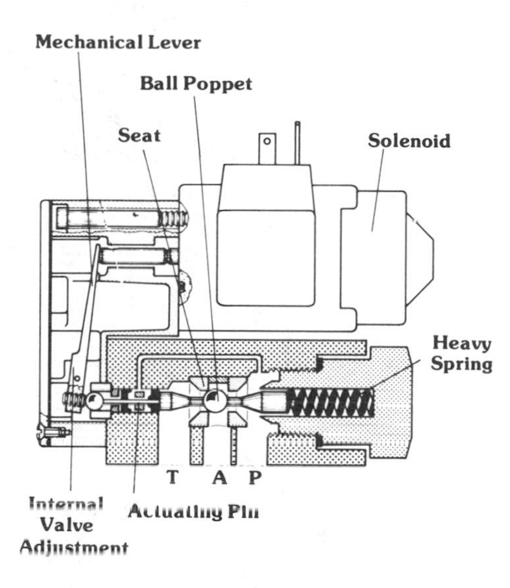

Figure 4.20 Single ball, normally-open three-way valve. (Courtesy of The Rexroth Corporation, Bethlehem, Pennsylvania.)

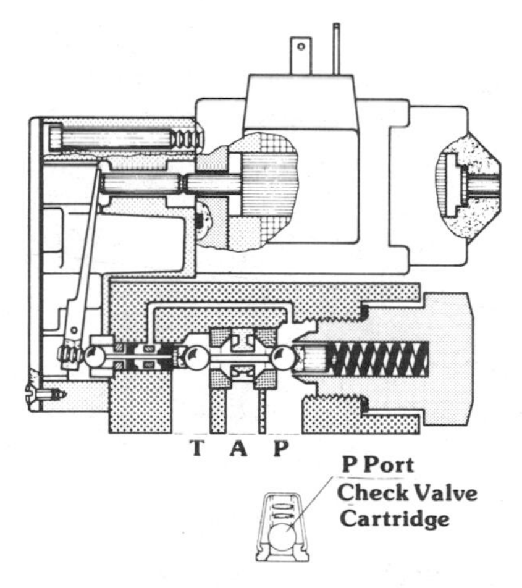

Figure 4.21 Double ball, normally-closed three-way valve. (Courtesy of The Rexroth Corporation, Bethlehem, Pennsylvania.)

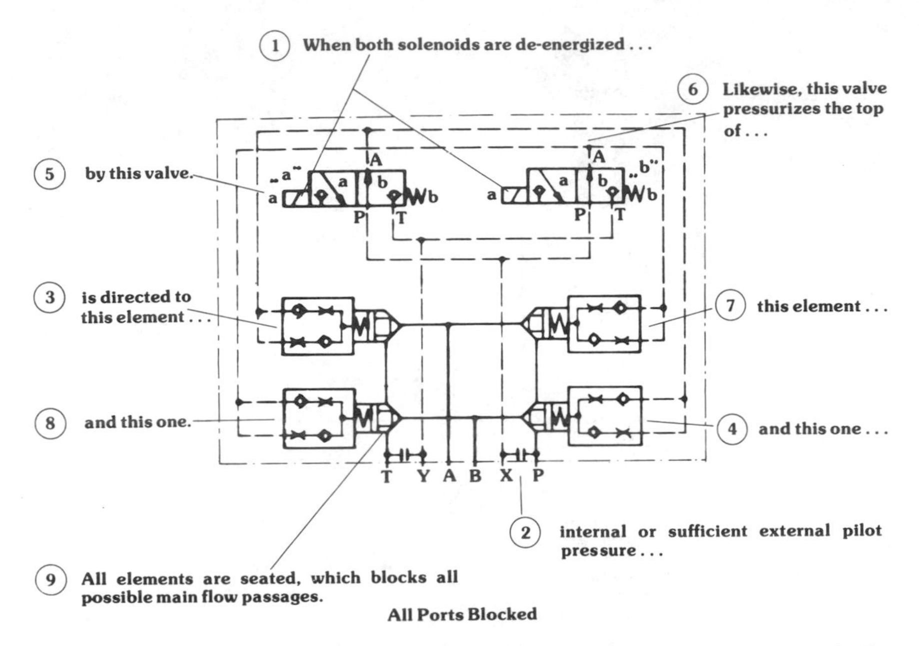

Figure 4.22 Logic circuit, both solenoids deenergized. (Courtesy of The Rexroth Corporation, Bethlehem, Pennsylvania.)

Threaded (O-ring-type with straight thread SAE-type configuration are preferred to taper-pipe thread) or flanged connection at the valve ports permit interconnection with pipe, tube, or hose. Many options are available in single-spool four-way directional control valves, particularly those operated manually.

Threaded, Flanged, or Manifolded Body – Multiple Spools To simplify plumbing and conserve space, many mobile systems use two or more valve spools in a single housing. Often the relief valve will be included in the assembly [Fig. 3.4 and 4.27(a)]. Usually one supply line and one return line to tank will provide the needed supply and return functions. The cylinder lines are run to the appropriate actuators.

Models can be provided with facilities to permit power *beyond* the initial valve assembly. Two or more valves can then operate in series. One pump is then able to supply both valves. Tank port from each valve is run independently. Power beyond port from upstream valve is connected to the pressure inlet of succeeding valves.

Valves specifically designed for mobile or marine service such as Fig. 4.27(a) often incorporate virtually all of the major control functions in a single assembly.

The valve assembly shown in Fig. 4.27(a) has a supply line at position 2. Relief valve (1) is pilot operated and externally adjustable. Discharge port (3) has facilities for series connection to another bank of valves (power beyond) with the capability to connect a return line for the upstream valve. In neutral position, the inlet (2) is connected to the outlet (3) via the series connections through the directional spools and valve coring within the valve housing (9). Cylinder ports (4) and (5) are either blocked in neutral or connected to tank according to the spool design. Spool C (13) is identified as a motor spool [Fig. 4.27(b)]. Both cylinder lines are connected together and to tank in the neutral or center position. Spool P (11) and S(12) are designed to be used with a single acting cylinder. Spool D (10) [Fig. 4.27(c)] is used for a double acting

Figure 4.23 Logic valve assembly. (Courtesy of The Rexroth Corporation, Bethlehem, Pennsylvania.)

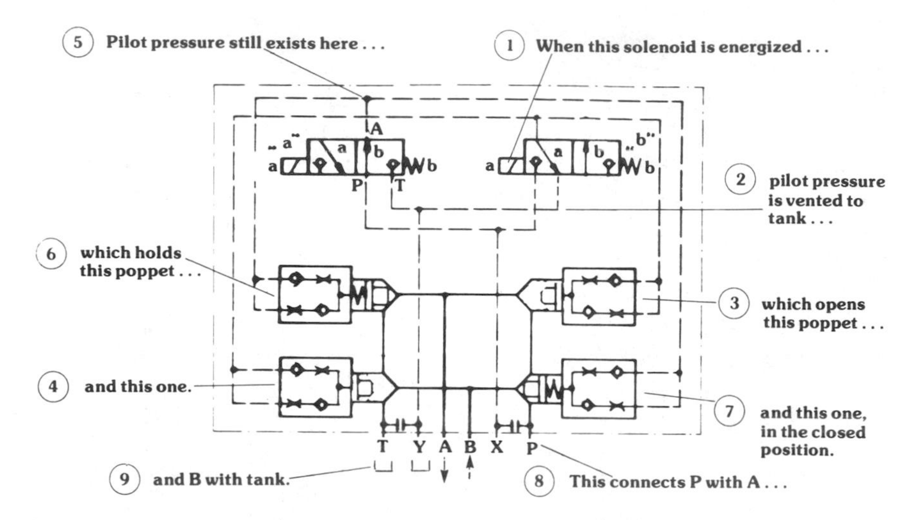

Figure 4.24 Logic circuit with P to A and B to T. (Courtesy of The Rexroth Corporation, Bethlehem, Pennsylvania.)

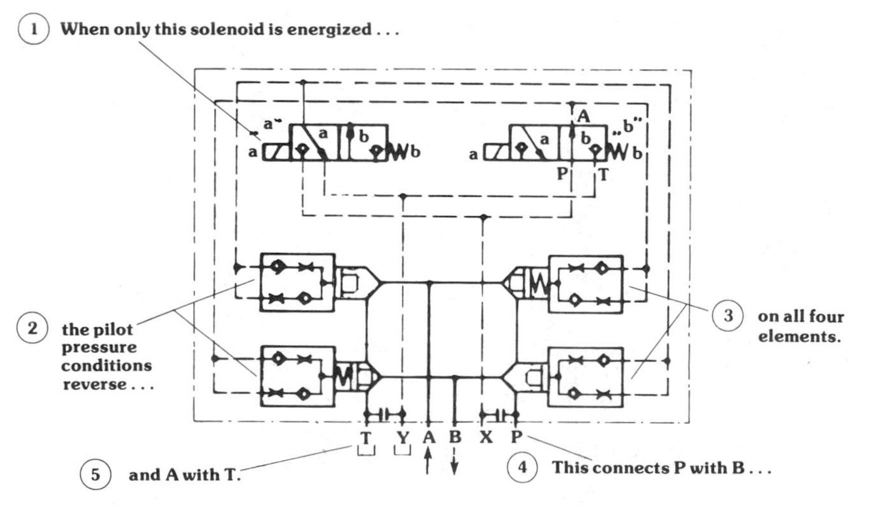

Figure 4.25 Logic circuit with P to B and A to T. (Courtesy of The Rexroth Corporation, Bethlehem, Pennsylvania.)

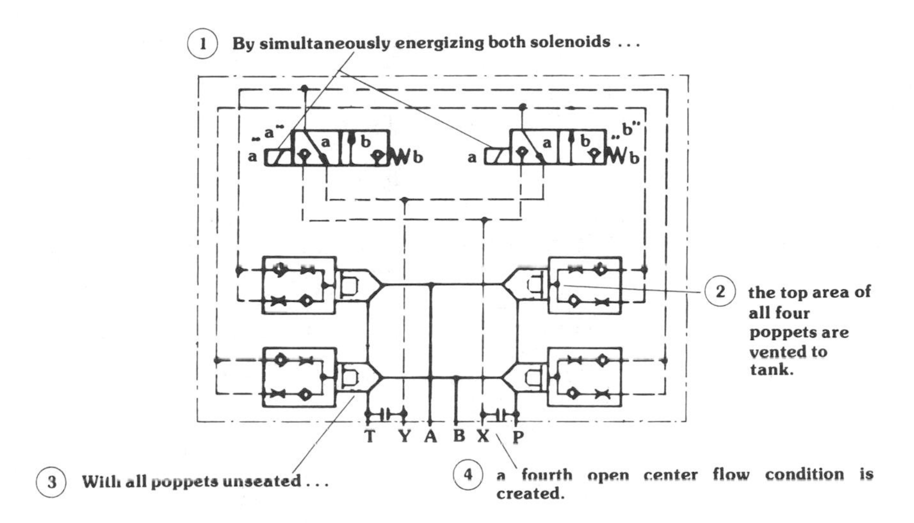

Figure 4.26 Logic circuit with both solenoids energized. (Courtesy of The Rexroth Corporation, Bethlehem, Pennsylvania.)

cylinder. A four-position spool [Fig. 4.27(d)] can be provided for operation equal to that of spool D (10). In addition, a float position has been added.

Check valves (6) [Fig. 4.27(b)] are inserted between the inlet pressure port and the supply to the cylinder to prevent actuator drop as the spool is shifted. These are known as *load checks.*

Metering notches (7) [Fig. 4.27(a)] provide better control of the actuator speed. Centering spring (8) can be replaced with mechanical detents, switch actuating mechanism, and/or automatic kickout devices as required by the machine function. Spring center mechanism provides a *dead-man* control so that actuator will stop or coast, according to spool center configuration, as operator releases the operating lever. A detent mechanism will hold position, such as taking up slack on a winch line, when operator chooses to operate other functions simultaneously without need for him to hold the detented spool. No action will occur unless the lever is moved to release the spool from the detent mechanism. An automatic kickout device can provide a safety or energy saving function by releasing the spool at a predetermined signal value. Crossover reliefs and make-up checks can be supplied internally to the ports of the valve leading to the actuator. Crossover reliefs will protect actuators when fluid is locked in the cylinder lines by the major flow directing spool. Make-up checks supply fluid to appropriate actuator lines when spool isolates these ports. A single rod cylinder actuator is used which involves a differential area or fluid passes to tank from the rotary motor drain as it coasts to a stop

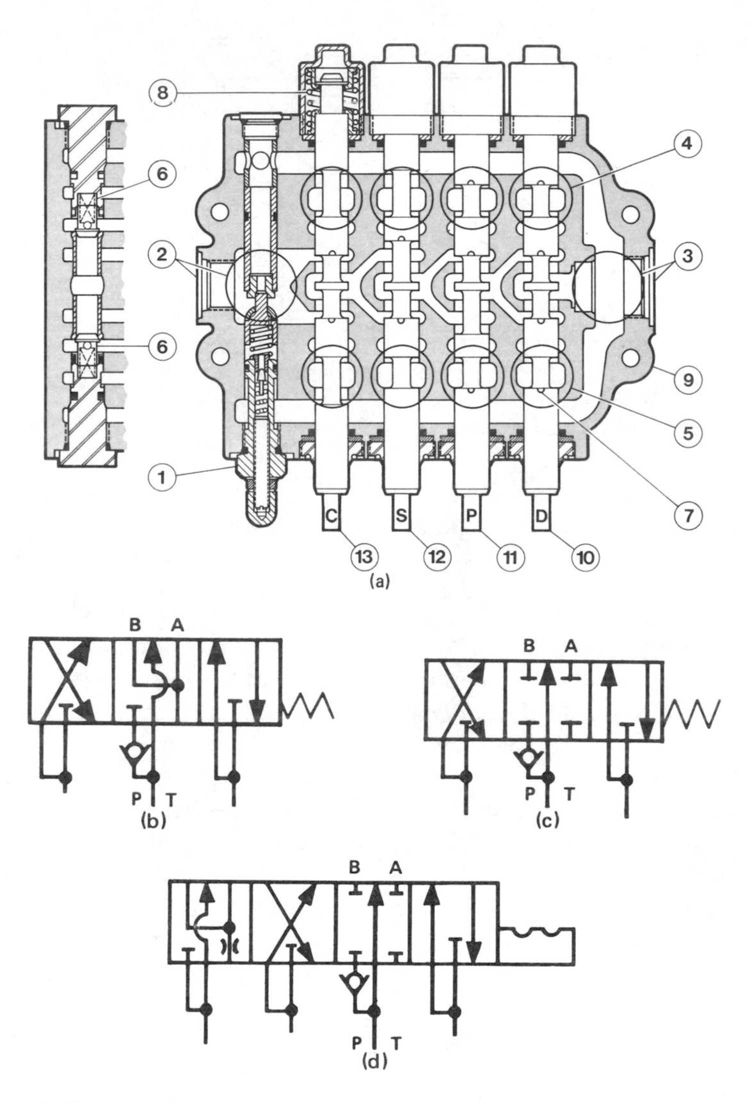

Figure 4.27 (a) Multiple spool monoblock control valve assembly. (b) Motor Spool Symbol. (c) Spool for double acting cylinders. (d) Four-position spool to include float position. (Courtesy of Parker Hannifin Corporation Mobile Hydraulic Division, Cleveland, Ohio.)

at the rate established by the crossover relief valve setting. Likewise, cylinder port relief valves (directing fluid to tank) can also be supplied.

The circuit as shown in Fig. 4.28(a) provides a parallel supply and parallel return to tank. Therefore, two or more spools can be engaged simultaneously.

With series pressure and parallel return to tank (Fig. 4.28(b)), the engaged spool closest to the valve inlet takes all pump flow.

Typical pressure drop versus flow curves for a six-spool valve are shown in Fig. 4.29. The exterior view of a six-spool valve assembly is also shown.

The monoblock structure of Fig. 4.27(a) can limit flexibility when mid-inlet connections are desired for special circuits. Modular valve structures usually have the front inlet module fitted with three ports; top inlet, side inlet, and side outlet. Unused ports are plugged. Mid-inlet modules are fitted with a top inlet port and provide a method to pipe output from a secondary pump source into a bank of valves. A high degree of flexibility of design is provided with the modular structure shown in Fig. 4.30(a). Practically all features available in the modular design. The exterior view of Fig. 4.30(b) shows the optional anticavitation checks located by the cylinder ports. These are used to reduce the chance of cavitation with an overrunning load. The optional overload relief valves just below the checks limit pressure in the specific cylinder line and, of course, in the circuit when the pressure is connected to that port. Spool assemblies in mobile valves can often be reversed for convenience in connecting to operating mechanisms. Spool seals can be externally replaceable. The individual load checks in the supply line prevent the load from dropping when changing spool positions and prevents backfilling from one cylinder port to another when operating two spools simultaneously. Spring to neutral can be replaced with detent positioning.

External ports are machined to accept SAE (Society of Automotive Engineers) standard straight thread fittings which provide leak-free connections with provisions to accurately position elbow-type fittings.

Figure 4.30(c) shows typical pressure drop from inlet to outlet ports with an open center sectional control valve. Pressure drop from the inlet to the furthest working module cylinder port is shown in Fig. 4.30(d). Pressure drop from the first working module cylinder port to tank is shown in Fig. 4.30(e).

Valves Mounted on a Subplate Subplate mounted valves [Fig. 4.31(b)] offer major advantages in the fabrication and maintenance of systems used in general manufacturing activities. A typical subplate (Fig. 4.32) can be connected with pipe, tube, or hose permanently without affecting the quick replacement of the control valve. Interconnecting of the valves can also be simplified by using a drilled, cast, or fabricated manifold. Virtually any form of fluid power control valve can be supplied to mount to a manifold plate or in cartridge form that can be threaded into a manifold structure.

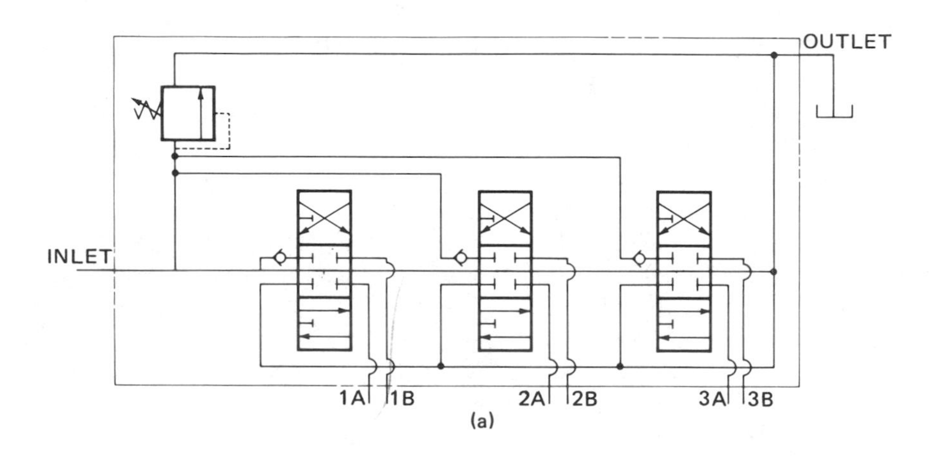

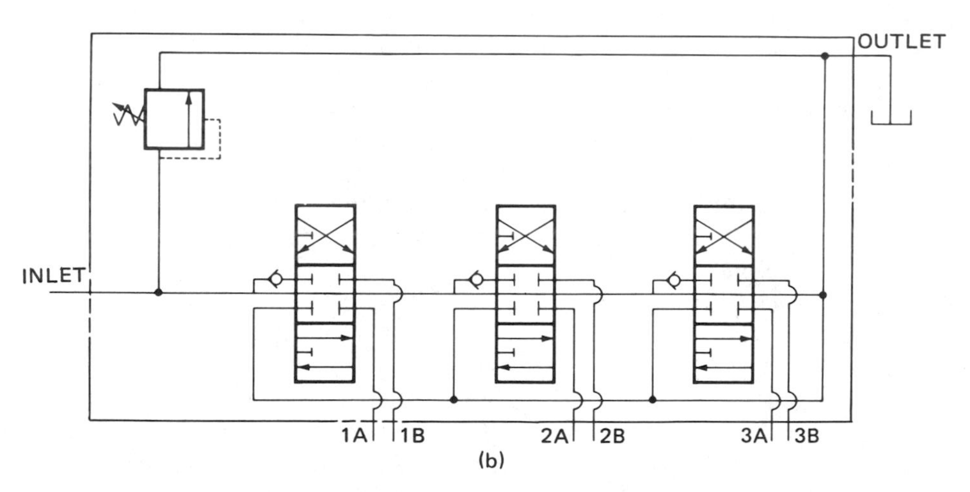

Figure 4.28 (a) Parallel supply and parallel return. Two or more spools. (b) Series pressure, parallel tank line. The engaged spool closest to the valve inlet takes all pump flow. (Courtesy of Parker Hannifin Corporation Mobile Hydraulic Division, Cleveland, Ohio.

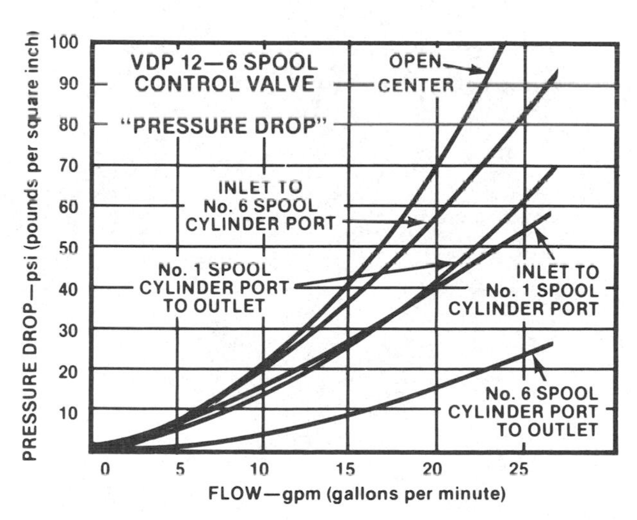

Figure 4.29 Typical six-spool valve with operating data. (Courtesy of Parker Hannifin Corporation Mobile Hydraulic Division, Cleveland, Ohio.)

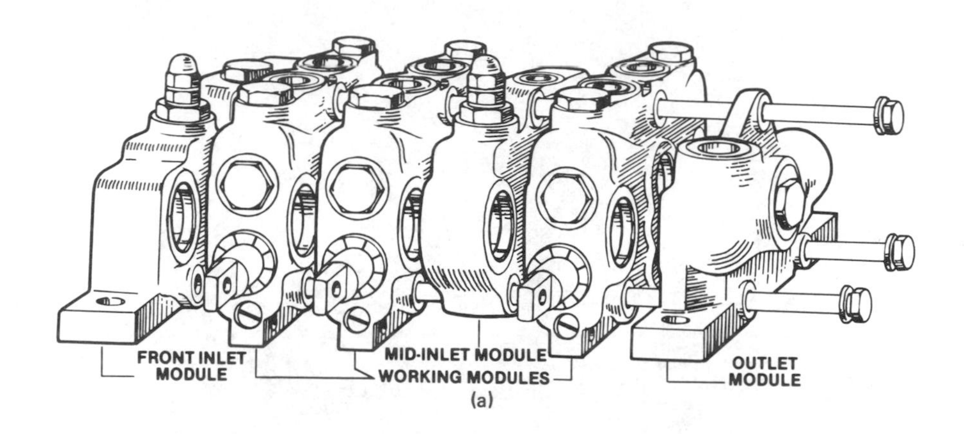

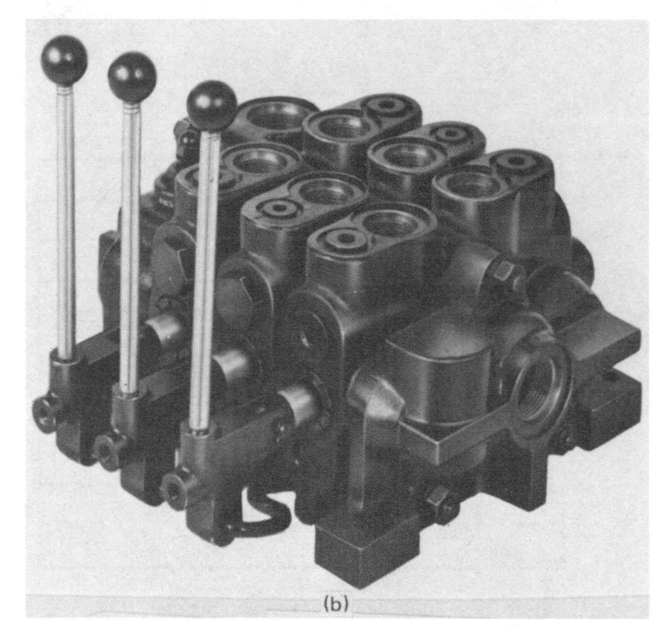

Figure 4.30 (a) Modular-type control valve assembly. (b) Lever-operated model with anticavitation checks. (Courtesy of Parker Hannifin Corporation Mobile Hydraulic Division, Cleveland, Ohio.)

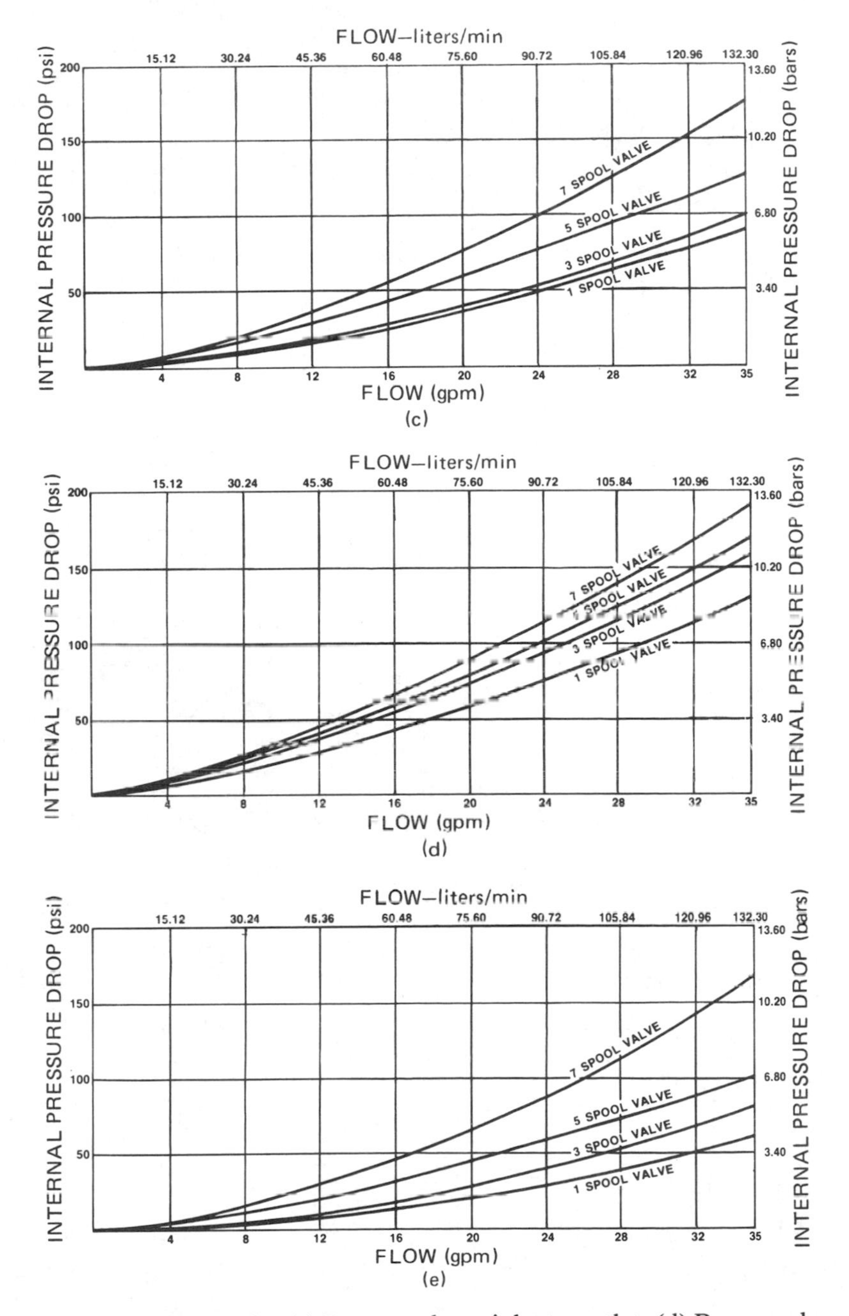

Figure 4.30 *(continued)* (c) Pressure drop, inlet to outlet. (d) Pressure drop, inlet to furthest working module port. (e) Pressure drop, first working module to tank. (Courtesy of Parker Hannifin Corporation Mobile Hydraulic Division, Cleveland, Ohio.)

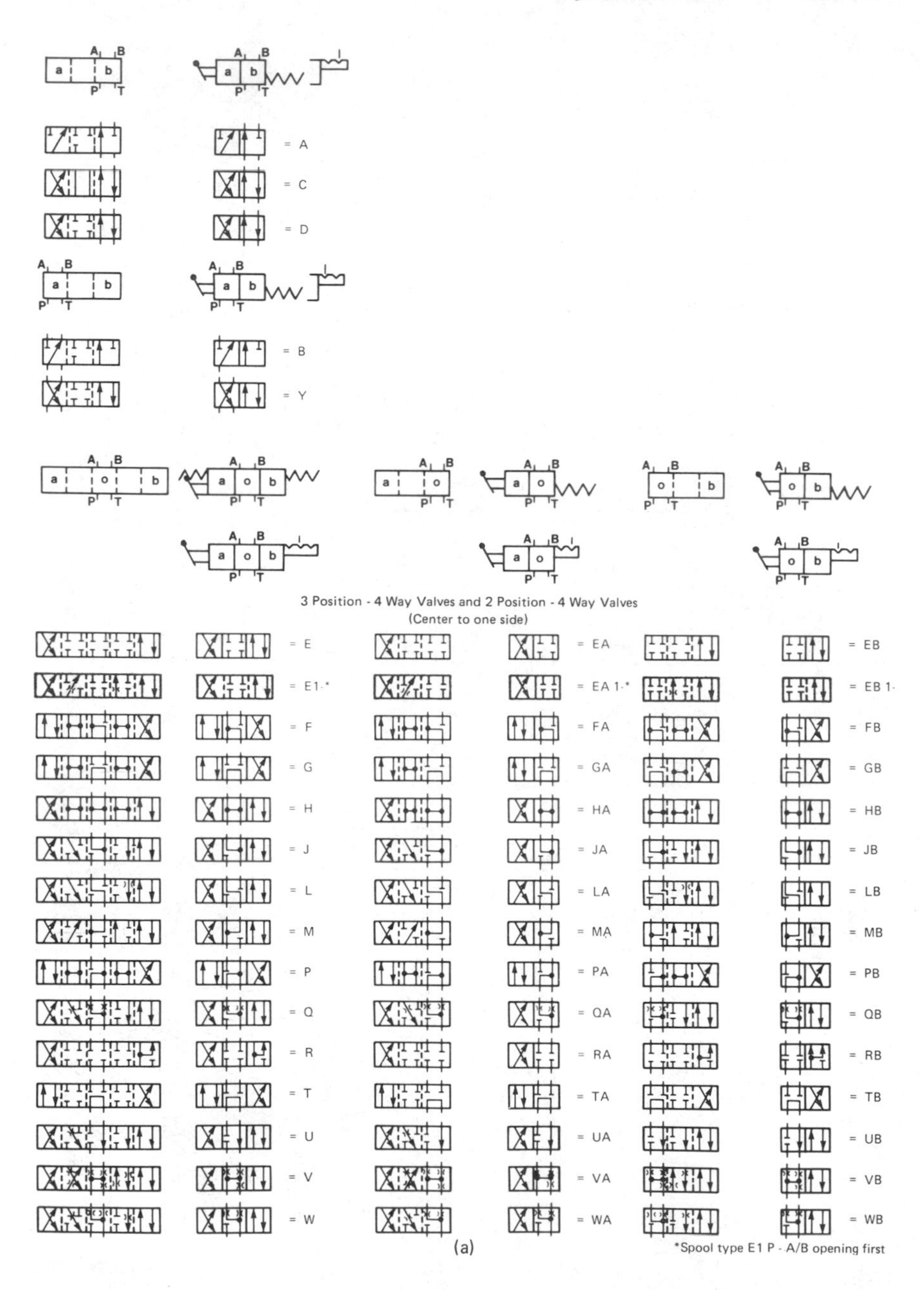

Figure 4.31 (a) Spool symbols. (Courtesy of The Rexroth Corporation, Bethlehem, Pennsylvania.)

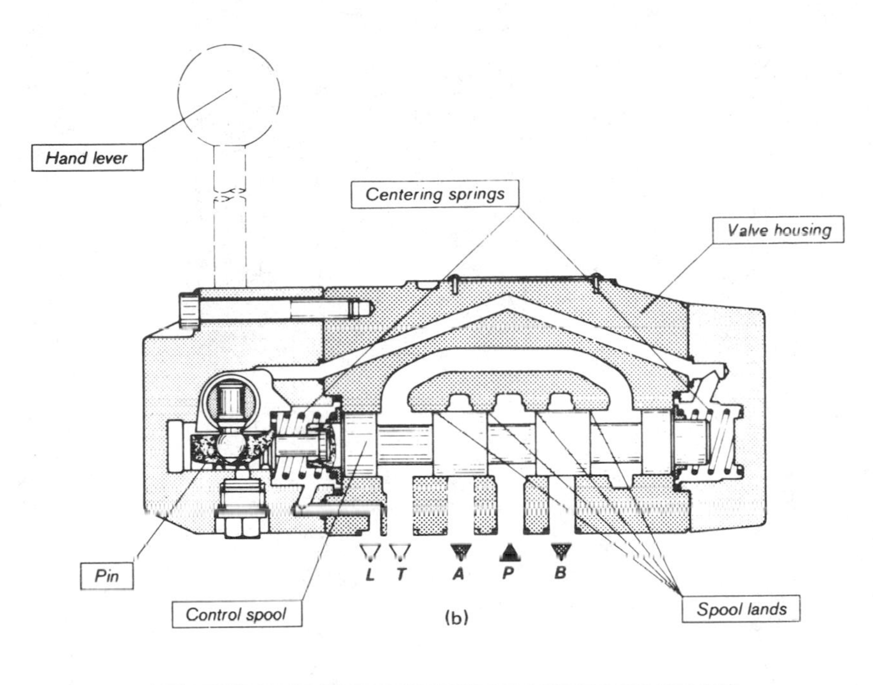

Weight	28.6 LBS	
Fluid temperature range	−20 to +160°F	
Viscosity range	35 to 1750 SSU	
Max. operating pressure (With tank pressure > 2400 PSI leakage oil must be separately drained)	Ports	
	A, B, P	T
	5000 PSI	3500 PSI

Hydraulic medium	Mineral oil		
Operating force	Detent	Spring offset or spring centred	
	18.7 LBS	24.2 LBS	
Operating angle	2 x 27,5° from center (see p.1)		
Throttling area in center position	spool type Q	spool type V	spool type W
	16 % of nominal area	16 % of nominal area	3 % of nominal area

(c)

Figure 4.31 *(continued)* (b) Lever-operated four-way valve. (c) Technical data for four-way valve. (Courtesy of The Rexroth Corporation, Bethlehem, Pennsylvania.)

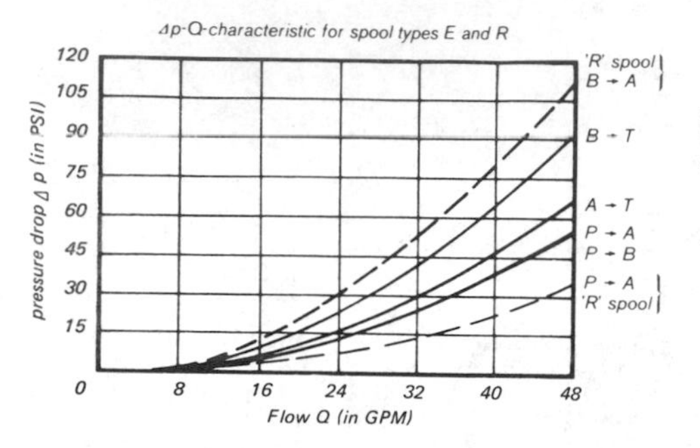

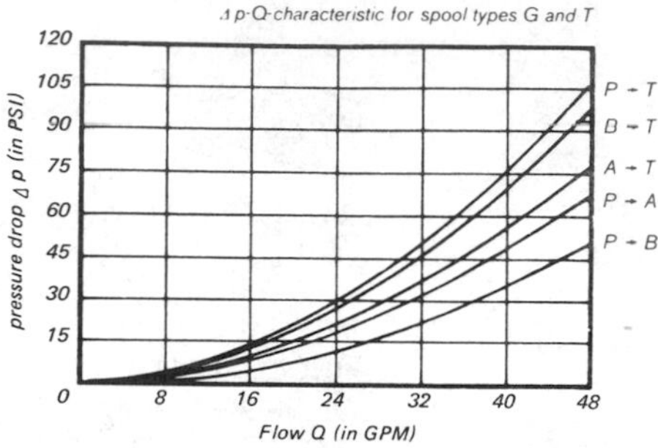

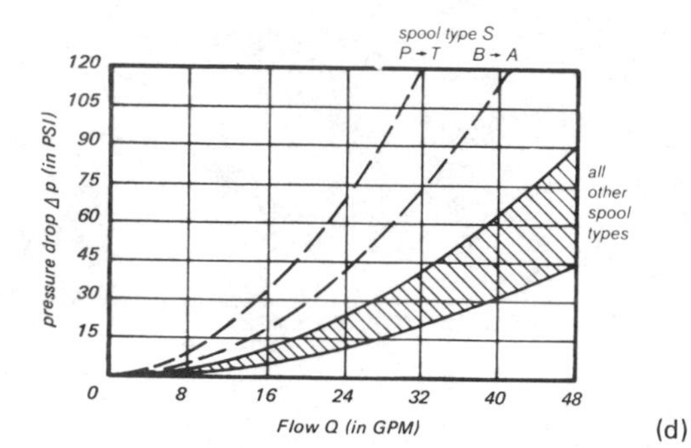

Due to silting the function of the valves is dependent on the filtration. In order to obtain the maximum flow values shown, main flow filtration of 25 μm is recommended. The internal flow forces in the valve also affect the flow, and therefore the flow details shown for 4-way valves apply for normal application with two flow directions (e.g. from P to A and simultaneously return flow from B to T).

(d)

Max. performance data	2-position valves, spring offset 3-position valves, spring centred				
Flow (GPM) for spool types	at pressure (PSI) of				
	1000	2000	3000	4000	5000
E, J, L, M, Q, R, U, V, W, C, D, K, Z	48	48	48	48	45
F	33	26	21	18	17
G, H, S, T	33	30	24	21	18
Max. performance data: 2- and 3-position valves with detent					
for all spool types	48	48	48	48	48

(e)

Figure 4.31 *(continued)* (d) Performance curves (Pressure drop Δp related to flow Q at 172 SSU). (e) General performance data. (Courtesy of The Rexroth Corporation, Bethlehem, Pennsylvania.)

Valves can be quickly replaced by draining the fluid from the manifold and removing the fasteners . When possible, manifolds are located above the tank so that valves can be replaced without need to lose fluid or drain the lines. If an overhead tank is used, then the lines to the subplate can be supplied with shutoff valves for servicing the equipment.

Downtime may be costly for high production machinery. The quick replacement capability may be a key to minimum production losses. In addition to quick replacement function, the subplate mounting minimizes strain caused by the basic plumbing connections. Often a dummy plate can be installed so that a new system can be thoroughly flushed prior to installation of the control valves. The valves may have been assembled in a sterile area and sealed in a suitable protective container for shipment and storage at the assembly site.

The directional control spool shown in Fig. 4.31(b) is actuated with a hand lever. The valve body is of a design that can accept end caps for pilot actuation either by a liquid or by compressed air. The valve body can also be fitted with a small pilot valve manifolded on the top surface actuated by solenoid(s) for directing pilot fluid to shift the main spool [Fig. 4.33(a)].

Figure 4.31(a) shows the various main spool combinations that can be fitted in a valve such as that shown in Fig. 4.31(b). Typical technical data for this type of four-way directional valve are shown in Fig. 4.31(c). The flow capabilities [Fig. 4.31(d)] of this specific valve are shown in typical Δp-Q-characteristics as related to specified spool type. Maximum performance values are charted in Fig. 4.31(e).

The typical subplate shown in Fig. 4.32 is designed and manufactured to an international standard that relates to the interface dimensions of the bottom mating surface of the valve and interface at the mounting surface of the plate or manifold structure.

Figure 4.33(a) illustrates a typical combination of the piloted main spool structure with a double-solenoid pilot valve fastened to the upper surface of the main valve.

The main valve with housing (1) is fitted with spring-centered control spool (2). The main spool could be fitted with three-position hydraulic pressure centering or with a single spring to bias the valve into a predetermined rest position.

The cavity in the main spool end cap (3) is connected with suitable passageways through the valve body to connect at the interface between the pilot valve and the piloted valve. One end cap is connected to the B port of the pilot valve and the other is connected to the A port of the pilot valve.

The interface between pilot valve (4) and the main body is designed to an international standard similar to that of Fig. 4.32. The solenoids (5) actuate the pilot spool. In the rest position the pilot spool is urged to a neutral position by the adjacent springs. Pilot pressure is blocked and pilot ports A and

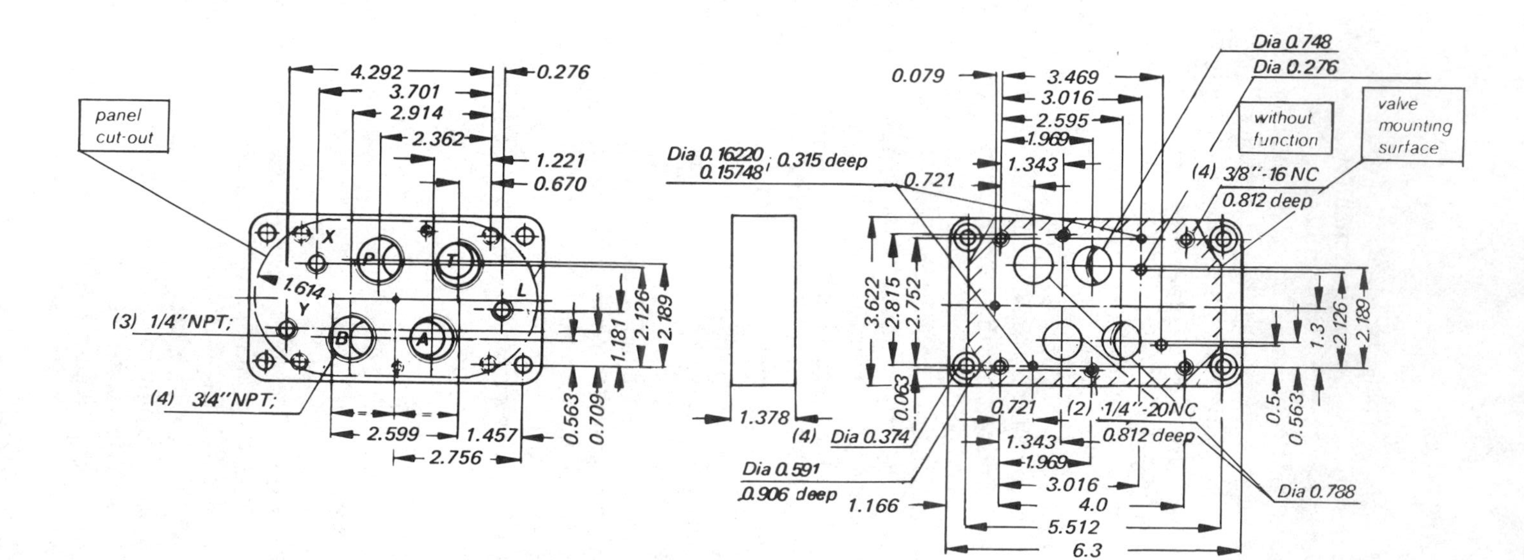

Figure 4.32 Typical ISO subplate interface. (Courtesy of The Rexroth Corporation, Bethlehem, Pennsylvania.)

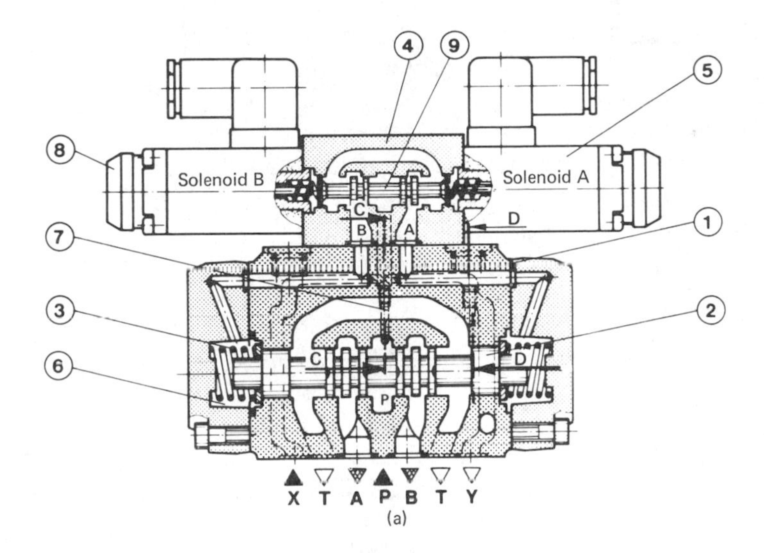

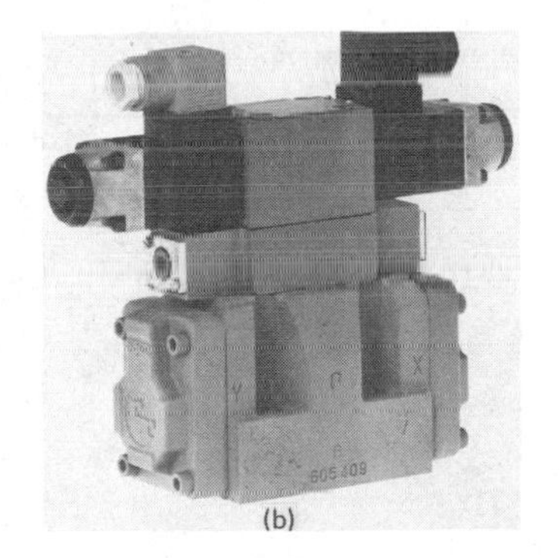

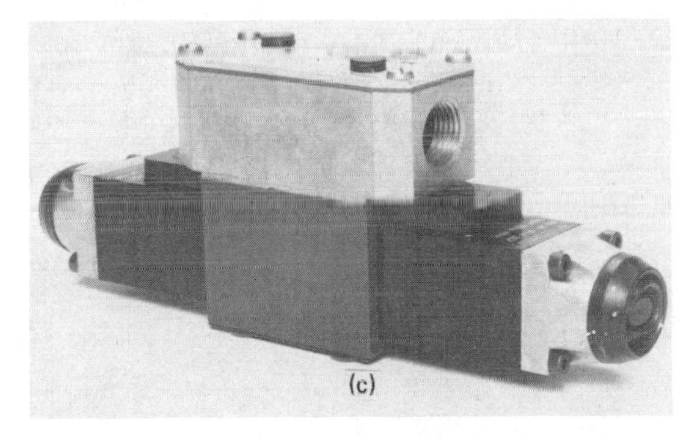

Figure 4.33 (a) Solenoid-controlled, pilot-operated four-way valve. (b) Pilot body with manual spool override provisions. (c) Pilot valve with signal lights. (Courtesy of The Rexroth Corporation, Bethlehem, Pennsylvania.)

B are directed to the tank port of the pilot valve, allowing fluid to escape from the end caps of the main body (6).

Pressurized pilot fluid can be supplied at the X port from a remote source or it can be taken from the internal drilling into the major pressure port by removing a threaded plug at (7). Some models have an O-ring sealed plug that can be reversed to change from internal to external pilot [Fig. 4.35(a)].

Drain fluid from the pilot valve and displaced fluid from the movement of the main spool can be returned to tank through the Y port or through an internal threaded hole at D. This hole is to be plugged if the Y port is connected.

Manual solenoid overrides (8) are available to allow the pilot control spool (9) to be moved without energizing the solenoid [Fig. 4.33(a)].

Figure 4.33(b) illustrates the solenoid-controlled, pilot-operated valve with separate electrical connecting terminals. A pilot valve with central terminal box and optional indicator lights is shown in Fig. 4.33(c). The optional lights indicate actuation of the solenoid and provide visual proof of electrical energy to the solenoid coil. The light is connected in parallel with the solenoid coil.

The valve shown in Fig. 4.34 is fitted with a sandwich structure between the pilot valve and the piloted valve. This structure incorporates a pair of flow control devices fitted with return flow checks. They can be installed to meter fluid to or from the main piloted spool cavities. These cavities control acceleration or deceleration of the end actuation.

The valve shown in Fig. 4.34 is similar to the valve in Fig. 4.33(a) except that pilot supply to the solenoid valve can be changed as per Fig. 4.35(a). The pilot choke adjustment (11) can be changed from meter-in (13) to meter-out (12) by removing the pilot valve, leaving the plate for the seal rings (21) in, inverting the pilot choke and replace, and replacing the pilot valve. Bolts (15) must be longer if a pilot choke assembly is incorporated in the valve assembly.

The pilot spool for a spring-centered main spool (Fig. 4.34) is designed

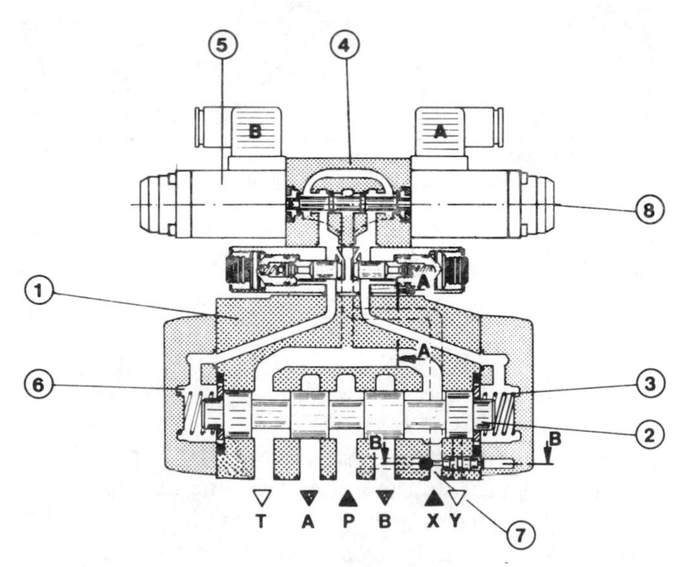

Figure 4.34 Solenoid-piloted valve with spool throttle controls. (Courtesy of The Rexroth Corporation, Bethlehem, Pennsylvania.)

to block pressure in neutral position and connect both main spool spring pockets to tank in neutral position.

A pressure-centered main spool [Fig. 4.35(b)] uses a pilot valve that blocks the return to tank port in neutral position and connects pilot pressure to both spring cavities adjacent to the end of the main spool in the relaxed position when no electrical signal is applied to the solenoids. Note the sleeve and piston assembly in the left end cap of the valve in Fig. 4.35(b). Pilot pressure directed to this assembly will allow the inner piston to shift the main spool against the centering spring. As the pilot valve is centered, pressure is applied to the major diameter of the spool, forcing the small diameter piston to the left until the large diameter sleeve is encountered at neutral spool position. Shift of the pilot spool to apply pressure to the major spool diameter and direct the flow to tank from the small piston and sleeve area will allow the major spool diameter to urge the spool to the left. This type of shift mechanism provides very positive spool positioning, responding to force values associated with pilot pressure rather than the minimal spring force available with usual centering spring.

Stackable Sandwich Valves Many control valve functions can be incorporated in sandwich plates to be inserted between the subplate and directional control valve.

Often cartridges can be inserted in the subplates. Additional function can be added in the sandwich plates. The obvious elimination of major interconnecting lines reduces first cost, saves potential leaks and loss of fluid, and insures long, satisfactory functional component life (Fig. 4.36).

Manifolded and Stackable Valves Sandwich, manifold, and stackable valves are available in many configurations. A careful study of circuit requirements will usually result in the choice of components most appropriate to the specific installation.

The valve assembly shown in Fig. 4.37 incorporates a pressure control valve, reducing valves to each directional control valve and suitable pilot structure. This particular assembly also has the capability to sense work loads and match input power needs very accurately to the work load. Appropriate shuttle valves are included to provide the load sensing signal in the proper sequence. Additional information is presented in Chapter 7.

4.7. MULTIPLE FLOW VALVES

4.7.1. Selector Valves

Multiple points within a hydraulic circuit can be interconnected by a valve such as that shown in Fig. 4.38 so that pressure can be sensed selectively, air can be drained, or other piloting functions can be accomplished.

Type 4 4 WEH 16..50/..6A..
This model has external pilot supply from a separate control line. Pilot drain is fed separately to tank via Y (external), not into the T line of the main valve.

9 Plug M6 DIN 906-8.8 SW 3
10 Plug M12 x 1.5 DIN 906-8.8 SW 5

Type 4 WEH 16..50/..6A..E..
This model has internal pilot supply from the P line of the main valve. Pilot drain is fed separately to tank via Y (external), not into the T line of the main valve.
Port "X" in the subplate is closed.

Conversion from external pilot supply to internal pilot supply.

Remove cover and bolt and refit the opposite way.

9 Plug M6 DIN 906-8.8 SW 3
10 Plug M12 x 1.5 DIN 906-8.8 SW 5

Type 4 WEH 16..50/..6A..ET..
This model has internal pilot supply from the P line of the main valve. Pilot drain is internal directly into the T line of the main valve. Ports "X" and "Y" in the subplate are closed.

10 Plug M12 x 1.5 DIN 906-8.8 SW 5

Type 4 WEH 16..50/..6A..T..
This model has external pilot supply from a separate control circuit. Pilot drain is internal directly into the T line of the main valve. Port "Y" in the subplate is closed.

10 Plug M12 x 1.5 DIN 906-8.8 SW 5

Pilot Supply, conversion internal/external.
Remove cover and pin and refit the opposite way.

Pilot Choke Adjustment
The pilot choke adjustment, designed as a sandwich plate, can be fitted between the pilot valve and the main valve. This is a double throttle and check valve (11).

The pilot supply or drain is throttled, depending on the mounting position of the pilot choke adjustment.

Clockwise rotation of the adjustment screw increases the shifting time of the valve, counterclockwise rotation decreases the shifting time.

Conversion From Meter-In to Meter-Out Control
Remove pilot valve, the plate for the seal rings remains, invert the pilot choke and replace, replace the pilot valve.

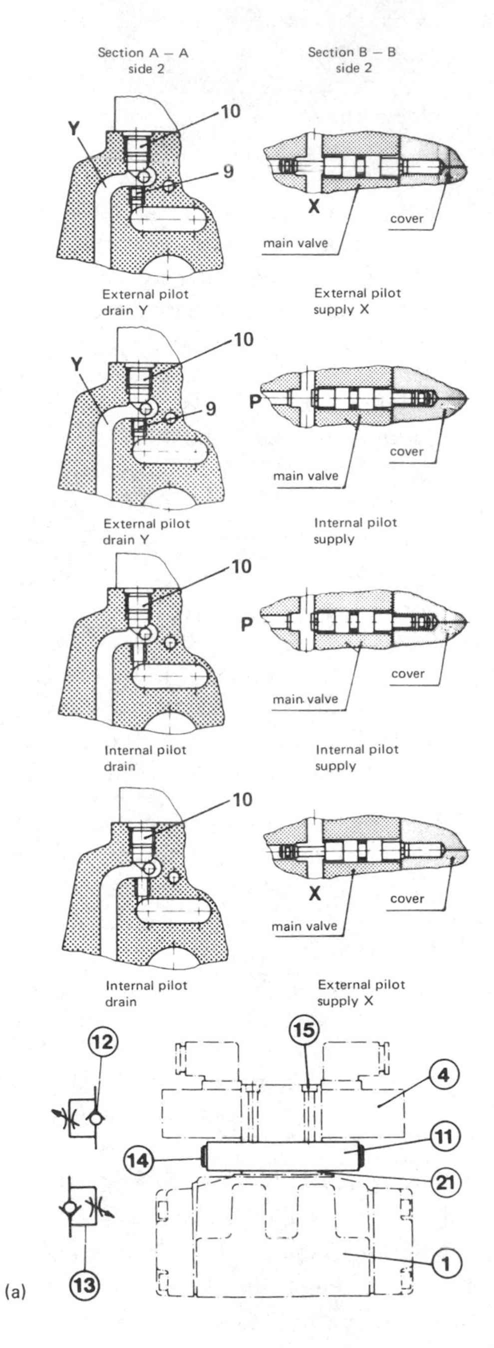

(a)

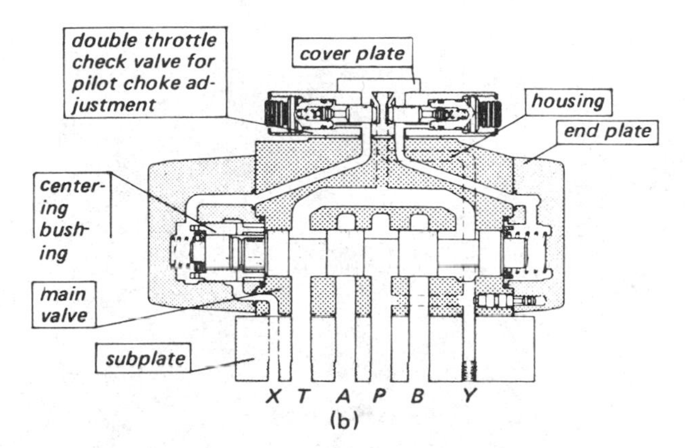

Figure 4.35 (a) Optional features in piloted valves. (b) Pressure-centered directional valve. (Courtesy of The Rexroth Corporation, Bethlehem, Pennsylvania.)

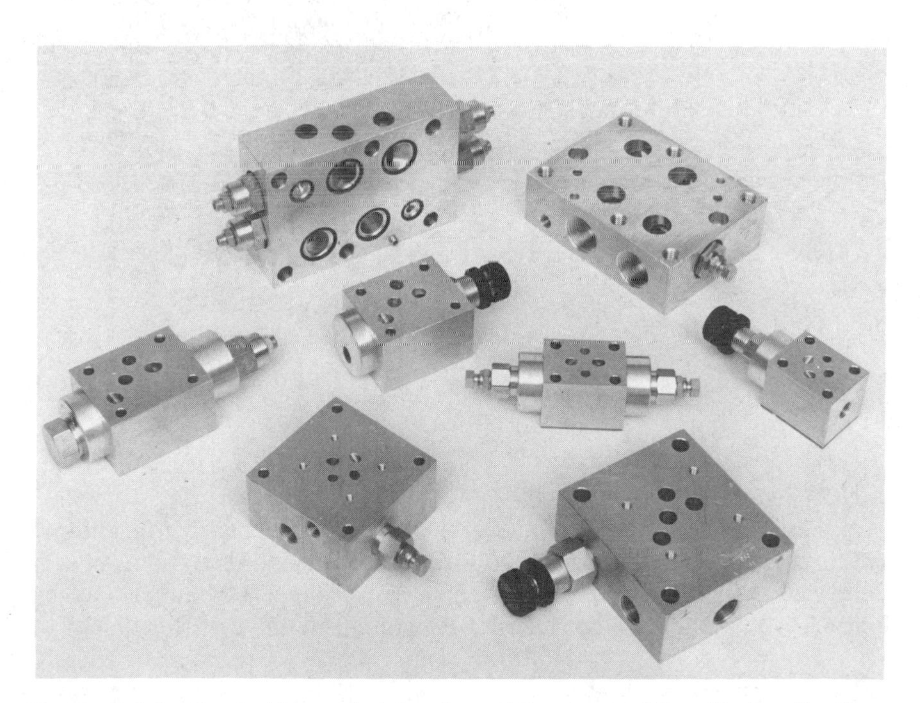

Figure 4.36 Stackable sandwich valves. (Courtesy of Sun Hydraulics Corporation, Sarasota, Florida.)

Figure 4.37 Subplate-type valve assembly. (Courtesy of HPI-Nichols, Sturtevant, Wisconsin.)

The valve shown in Fig. 4.38(c) symbolically indicates the facilities to selectively connect to six specific points. The ports not directly connected to the key port can be blocked in one configuration or drained to tank in another. Fig. 4.38(b) shows the basic internal structure of a valve of this type.

4.7.2. Separate Tank Metering

Directional valves may be modified so that pressurized fluid is directed to one cylinder port just prior to allowing oil to return to tank through a second port, thereby preventing cavitation within a cylinder or hydraulic motor.

(a)

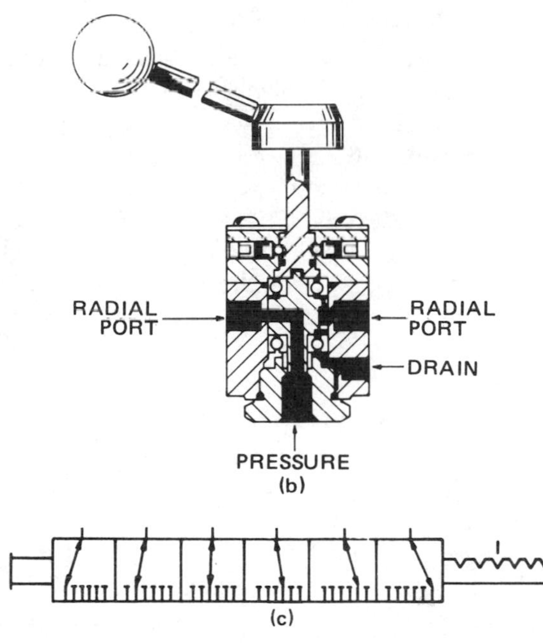

(b)

(c)

Figure 4.38 (a) Selector valve. (b) Cross section vie of selector valve. (c) Symbol for selector valve. (Courtesy of Rexnord Fluid Power Division, Racine, Wisconsin.)

4.7.3. Providing Auxiliary Capacity

Pilot-operated two-way valves may be used in parallel with certain ports to add return to tank capacity without major increase in the size of the basic four-way directional valve. Note valve B in Fig. 3.30

4.8. VALVE ACTUATORS

4.8.1. Lever, Pedal, and Treadle

Operator actuated valves are quite common in mobile and marine equipment; visual response to machine activations is usually of major significance.

4.8.2. Cam Operation of Directional Valve

Two-way cam-operated valves such as that shown in Fig. 4.11 respond to a preprogrammed mechanical cam that insures repetitive accurate flow rate, acceleration, and deceleration patterns. Change in pattern requires a physical change of the cam structure. Such units are generally used in dedicated purpose, mass production, machines.

4.8.3. Solenoid and Motor-Operated Valves

An electric signal to a hydraulic valve can be accepted in a simple solenoid structure to physically move a valve member in a strictly digital on-off mode. An electric motor device with a gear train can operate a cam or screw device to actuate the valve member in a somewhat less harsh manner.

A direct current solenoid as employed in the valve of Fig. 4.37 can move the valve spool in a pattern directly responsive to the magnitude of the signal strength and associated voltage level. This pilot valve movement can be amplified hydraulically to accurately position the major flow directing spool which controls both direction and rate of flow through the valve assembly.

4.8.4. Piloted Valves

Hydraulic – Applied or Released

The pilot valve of Fig. 4.37 directs pressurized fluid to the major directional (slave) spool. The majority of pilot control functions deal with the application of pilot pressure.

The pilot of Figs. 3.10 and 3.11 function by releasing pilot pressure. The option of releasing pilot pressure can simplify the control systems because the pilot pressure can often be provided internally through a suitable orifice. The small loss of fluid entailed in the pilot function may be insignificant in the overall operation of the transmission system.

Pneumatic Pilot Control

Both hydraulic pressure and directional controls are available with facilities for piloting with pneumatic signals. The purpose for using air may be that of economy when compressed air is readily available and hydraulic pressure for pilot purposes may require significant investment and/or maintenance costs.

Direct operation of some units with instrument air in the 15 psi area where clean air is available provides excellent response characteristics.

Usually pilot air must be clean and dry to provide needed control functions.

4.9. PANEL ASSEMBLIES

Many hydraulic assemblies are used in mass produced machines. It becomes economically feasible to incorporate many controls in one block as shown in Fig. 4.39

Thc control system shown in Fig. 4.37 also contains the major pressure, direction, and flow controls in a complete hydraulic valving system for a machine. Obviously, some valves may be best located in the cap of a cylinder or at a point where flow can be directed with minimum energy loss.

Figure 4.39 Custom valve assemblies. (Courtesy of Sun Hydraulics Corporation, Sarasota, Florida.)

Economics, maintainability, and dependability may be the criteria upon which design parameters are based.

4.10. SUMMARY

Directional control valves may encompass some areas of pressure and flow control as well as the basic directional control function.

Choice of envelope design is usually dictated by end use expectations. Envelope and connecting devices can be lightweight for airborne service, rust resistant for marine service, and easily serviced for installation on mass production machinery. Dependability, cost, maintainability, and projected life also hinge on the expected end use.

Many systems are operated in areas where maintenance is virtually impossible. Under such circumstances, some components are considered throwaway items and ease of complete replacement of the suspected component is a key factor.

5 Flow Control Valves

5.1. INTRODUCTION

The control of fluid *flow rate* in a hydraulic power transmission system can be accomplished in three basic ways.

5.1.1. Variable Delivery Pumps

Perhaps the most efficient control of flow rate is by use of a variable delivery pump which can be controlled by manual or automatic means to deliver only the flow rate needed to accomplish the assigned task at the desired rate of speed and at a pressure value appropriate to the work being performed. The major losses in such a system are associated with friction, both mechanical and that resulting from the flow of liquid through the conductors.

A major disadvantage to such a system is the limitation of work output to one actuator. As a result of this limitation, such *hydrostatic transmissions* find major usage as traction drives and like applications.

In summary, the control of rate of flow with a variable delivery pump usually uses no restrictive or diversion-type valves. The control of rate of flow results from a change in the pumping mechanism in response to a mechanical or automatic signal from a control source.

5.1.2. Restrictive Devices

Pressurized fluid metered into that portion of a circuit controlling the actuator delivering linear or rotary force and motion will establish the rate of movement. Flow at an increased rate will increase the rate of movement of the actuator output if within the force capabilities of the actuator. Conversely, a reduced rate of flow will cause a decrease of the rate of movement of the actuator output.

The fluid can be metered into or out of the actuator. Circuit characteristics will determine which method is best for the specific power transmission task.

Some considerations that determine the decision to meter fluid into or from the actuator are predicated on the type of control needed and the physical characteristics of the actuator.

Metering fluid from the actuator provides a restrictive action that can result in rigid control of tools such as drills and milling cutters that often tend to *pull ahead* by the cutting action with danger to the tool or work. Metering-out prevents external forces from affecting the rate of movement. Metering fluid to the actuator may be required if there is the potential for intensification of the fluid because of the differential of areas between the piston face and the rod diameter of a hydraulic cylinder. As an example, a cylinder with a piston rod with an area one-half that of the piston face is common. If fluid is metered from the rod end cavity, we can expect a 2 : 1 area ratio to result in the development of a pressure of 6000 psi at the rod end cavity if we provide 3000 psi at the piston face. This is usually an unacceptable condition and potentially dangerous to machinery and personnel.

A compromise can be reached by the use of meter-in of fluid to the piston face area and meter-out through a counterbalance or holding valve at the rod end of the cylinder with a pilot signal from the head end connecting line between the outlet of the metering valve and the cylinder face area. Rotary actuators may require an external drain to tank to protect the shaft seal if fluid is metered-out and both ports are pressurized.

Some rotary actuators have a check valve network that drains the shaft seal to the lowest pressure port connected to the actuator housing. If both pressurized ports are at a maximum pressure value because of the flow control restricter then the resulting pressure can be impressed on the seal. Potential rapid wear and/or extrusion of the seal may occur. An external drain for the seal cavity can eliminate this potential problem. Certain actuator designs anticipate this problem and provide seals designed for the highest expected pressure. This should be determined prior to the choice of metering-in or metering-out from the actuator, or if an external drain is needed. When the fluid is metered it implies a restrictive action. Fluid not passing through the orifice or metering structure must be addressed. A variable delivery pump can be automatically adjusted to pump sufficient fluid to maintain a predetermined pressure level. Thus the pump responds to the rate of flow through the restrictor or multiple restrictors in a parallel circuit.

A constant flow pump circuit incorporates a pressure level control valve which diverts excess fluid (usually to tank) at a predetermined pressure level.

The fixed delivery pump must supply more fluid than that being metered-in or out of the actuator circuit.

5.1.3. Bypass or Bleedoff Devices

The rate of flow from a constant flow pump may be too much to provide the desired output speed from an actuator. By diverting some portion of the flow to another area or to the tank, the desired speed can be attained.

In a similar manner the first rate of actuator movement to be considered can be established and the excess flow can be diverted to a second circuit. This is a priority circuit wherein the first circuit must be satisfied prior to diversion of pressurized fluid to the downstream or secondary area.

In the bypass, diversion, or priority function, the pressure level is established by the load on the primary actuator until it is satisfied rather than excess fluid being passed over a relief valve at maximum set pressure.

In summary, the hydrostatic transmission using a variable delivery pump to control rate of flow is not within the scope of the subject matter of this volume. Valves to control rate of flow by restricting flow are appropriate to this presentation.

5.2. RESTRICTIVE DEVICES

5.2.1. Noncompensated Restrictions

A restriction to flow of a finite size will pass fluid at a specific rate if upstream fluid pressure and temperature remain constant. Some heat will be generated at the point of restriction as work is performed in the fluid movement through the restriction. A change in pressure will change the flow rate through an orifice. A change in temperature will affect fluid characteristics so that a change in flow rate will also occur.

The magnitude of the change that will occur with a change in pressure or temperature may be of little consequence in many circuits. Because a minimal change in rate of flow can be tolerated in many circuits, the restrictive device can be quite simple.

Fixed Orifices

A fixed orifice may be the least expensive device to regulate the rate of flow from one part of a hydraulic circuit to another. Flow characteristics of a fixed orifice are related to the length of the orifice and the diameter. As the length of the orifice is increased, it also increases the rate of change of flow resulting from changes in pressure and/or temperature. The term *sharp-edged orifice* indicates an orifice with only sufficient length to prevent the disintegration of the structural material from the pressure differential or erosion. The sharp-edged orifice is least affected by change in temperature and/or pressure.

An orifice can be drilled in a removable plug or in a plate so that the size can be changed without materially altering the valve or piping. A temporary hole size reduction can be effected by inserting a pin or piece of wire in the drilled hole, secured so that the fluid flow will not dislodge it. The diameter of an orifice, and sometimes its length, may be noted on a drawing adjacent to the orifice symbol as shown in Fig. 5.1. It may be shown in metric and/or inch values which for purposes of clarity must be identified. Standardized orifice plugs, plates, and pipe fittings are available from several sources.

An orifice in a pipe, tube, or hose fitting may be calibrated so that it will pass normal working flows. If a pipe or hose line were to rupture or be torn from the machine while in service, the orifice fitting (usually located in a cylinder cap or motor housing) will prevent an excessive rate of movement and minimize potential danger to personnel or machinery.

Needle Valves

Change in the size of an orifice can be important to many circuit functions. A needle-type shutoff valve such as that shown in Figs. 4.11 and 4.12(a) can be used as an adjustable orifice as well as a two-way shutoff valve. The cartridge-type needle valve (Fig. 5.2) has as a primary function that of an adjustable orifice.

Needle valves are available with modifications intended to provide sensitivity of adjustment appropriate to this function as an adjustable orifice.

Spools and Spigots

Spools and spigots are rarely used as a restrictive flow control device without some form of compensation to insure a constant pressure drop across the resulting orifice.

5.2.2. Compensated Restrictions

The term compensation as associated with a flow regulating valve indicates that some device has been incorporated in the basic valve to insure a constant flow through the orifice regardless of upstream or downstream pressure values if operated within the design parameters of the valve. *Temperature compen-*

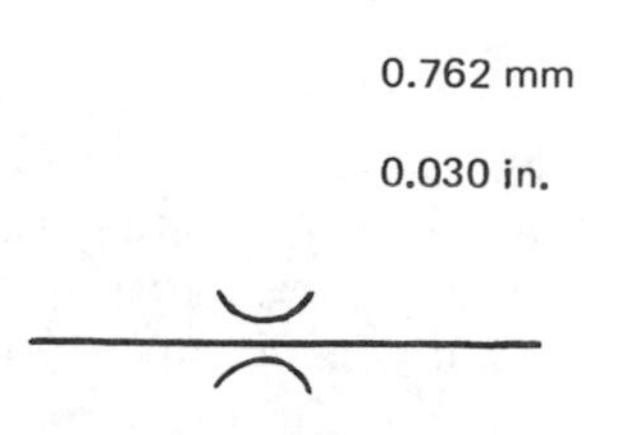

Figure 5.1 Orifice size shown adjacent to symbol.

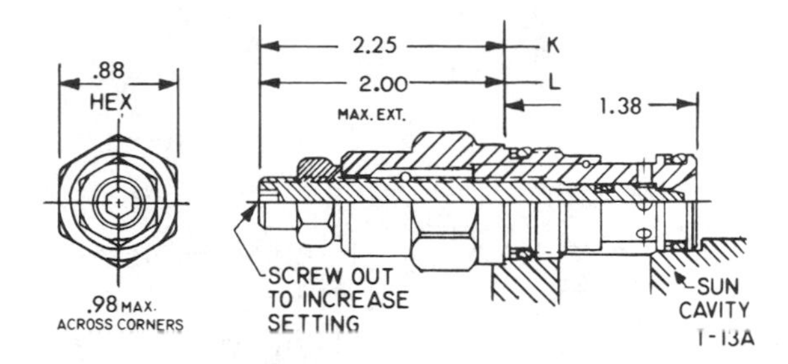

Figure 5.2 Cartridge-type needle valve. (Courtesy of Sun Hydraualics Corporation, Sarasota, Florida.)

sation may be by use of a sharp-edged orifice or introduction in the orifice structure of a metal device that will expand or contract with change in temperature at a rate that will automatically adjust the orifice in the proper relationship.

Reducing Valve Compensation

The reducing valve of Fig. 3.15 can maintain a constant pressure at the low pressure outlet port. If the downstream side of the restriction were to tank, there would be adequate compensation to maintain a uniform flow regardless of the pressure value ahead of the reducing valve.

Many flow control valves have ranges or varying pressures at the outlet so that facilities are needed to integrate the resulting information in the valve structure and automatically adjust for the downstream resistance to flow.

Threaded Body Figure 5.3 shows a nonadjustable compensated flow control valve. Fluid entering the inlet port impinges on the full face of the spool, urging it against the spring. The spool can move against the spring until it closes off the outlet port. Then a flow is established through the fixed orifice in the spool. This flow equalizes the pressure on both ends of the spool. The spool is then in balance. This movement opens the outlet port. A balance will automatically be established between the amount of fluid that will pass through the orifice and the upper edge of the spool. Note that the open end of the spool is attempting to shear off the flow of fluid to the outlet port. The flow rate through the fixed orifice is determined by the size of the hole and the value of the spring loading. The heavier the spring for a given-size orifice, the greater the fluid flow; the larger the hole with a fixed spring value, the greater the fluid flow. When a spring size is established and an orifice size is chosen, the repetitive flow will remain the same regardless of the pressure upstream or downstream because the spring and orifice size determine the flow. The area of the spool is equal at both ends; no differential is creat-

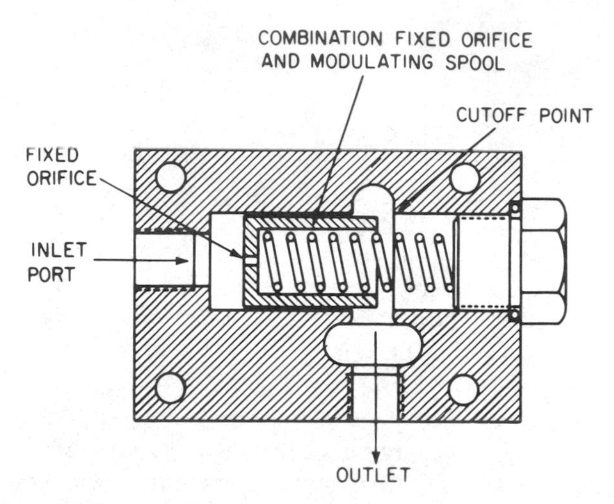

Figure 5.3 Nonadjustable compensated flow control valve. (From Pippenger, John J., and Hicks, Tyler G., *Industrial Hydraulics,* 3rd Edition, Gregg Division, McGraw-Hill Book Company, New York, New York, 1979.)

ed. The spool may be located within a body with various port configurations. As an example, a return flow through the body may be provided. This spool mechanism may also be used in other valve structures to provide a compensated-orifice effect.

A valve of the above design is used to control the lowering speed of pallets on lift trucks and similar devices where the pressure values may be different during each functional operation. The compensated flow control using the design in Fig. 5.3 must have a certain mimimum flow of fluid to function so that the orifice will create enough restriction to move the poppet. The pressure required to move the poppet is usually a relatively low value and does not hinder effective operation of conventional machinery.

Suplate Mounted Figure 5.4 shows an adjustable compensated flow control valve. The cross section of the valve shows its construction, and the symbolic drawing at the right shows the principle of operation. The symbolic drawing shows the inlet connected directly to a reducing valve. A signal to actuate the normally-open spool in the reducing valve is taken from the downstream side, just ahead of the adjustable-orifice. Another signal is taken downstream from the adjustable orifice. This signal is reflected back into the spring cavity of the reducing valve.

Assume that the spring in the reducing valve can be compressed to a point where flow through the valve will cease at 150 psi pressure at the orifice. Then, when the adjustable orifice is completely closed, there will be no pressure above 150 psi at the input of the orifice. This assumes that no pressure is reflected into the spring chamber of the reducing valve. Now, as long as there is a supply of fluid under pressure of more than 150 psi at the inlet port and a volume of fluid exceeding that to be controlled by the valve, the pressure drop across the orifice will always be 150 psi. The reducing spool will control the pressure at the orifice because the signal to the reducing spool is taken from the input to the adjustable orifice.

When a load is handled by metering fluid to a cylinder or other device, there will be fluid pressure on the outlet of the flow valve. To compensate for this pressure, a control line is provided from the flow valve outlet back into the spring chamber of the reducing section. This line senses the value of the load and adds it to the bias spring, thus urging the spool to the open position. Thus, the pressure signal to one end of the reducing spool is taken from ahead of the orifice, and the energy available to bias the spool is made up of the spring and the pressure sensed beyond the orifice. In a typical installation, the load consists of a dead weight lifted by an elevator mechanism. Two cylinders functioning together are required. They must lift with an equal rate of travel to avoid twisting the product. A compensated flow control valve at each cylinder used for ejection will ensure a rate of movement sufficiently accurate to prevent the product from skewing during the ejection portion of the cycle and will keep both ends traveling at nearly equal speed. In this installation the input or pressure is not necessarily stable, but it is always ade-

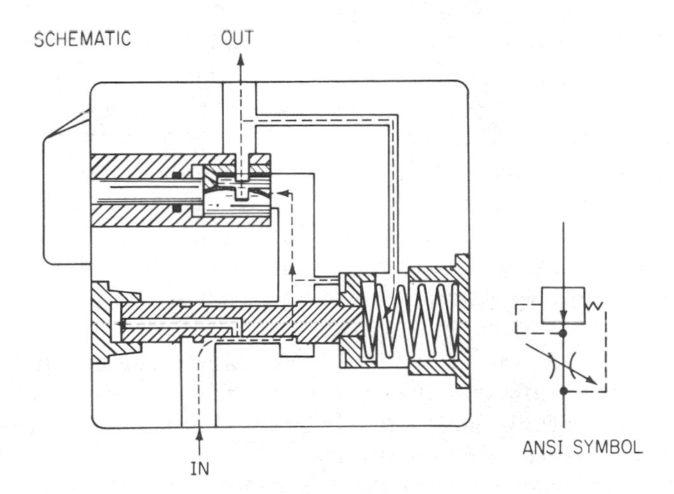

Figure 5.4 Adjustable compensated flow control valve and symbol. (Courtesy of Rexnord, Fluid Power Division, Racine, Wisconsin.*)

*Now a division of Dana Corp.

quate for the ejection operation. The weight of the part being lifted is constant, but the friction encountered may vary. Variations in input pressure and flow volume are caused by the actuation of several other parts of the machine concurrently with the ejection device. The variable inlet pressure is compensated for by the reducing valve. The variation of outlet load is reflected back to the spring chamber of the reducing valve where it is compensated as described above.

Figure 5.4 shows that fluid entering the valve inlet is directed to a two-diameter reducing spool. Two diameters are used to provide a precise control at the cutoff point and to provide the desired area to receive the operating signals for the reducing valve. The areas at the bottom of the spool and under the head opposite the spring cavity are ported to the outlet of the reducing valve prior to entering the orifice. The spring cavity is connected to the outlet of the flow control valve. In this valve the adjustable orifice is a spigot that varies the flow by use of a contoured notch over an arc of approximately 180°. A drain line is not necessary because of the use of the high-pressure, captive, O-ring-type seal on the adjusting shaft. Basic construction may vary with different manufacturers.

The valve of Fig. 5.5(a) incorporates internal pressure balancing for easy adjustment of flow setting even when fully pressurized. Pressure compensation will maintain a preset flow within 1% to 5%, depending on the basic flow rate as long as there is 150 psi pressure differential between the inlet and the outlet ports.

Figure 5.5(b) shows the typical dial setting versus flow for a pressure and temperature compensated flow control valve. Note that the dial is provided with 24° rotation per increment. The temperature compensation that can be expected in a typical installation is shown in Fig. 5.5(c).

If the inlet pressure to a valve such as that shown in Fig. 5.5(a) is 3000 psi, there will be a slight change in flow across the orifice as the outlet pressure is varied. The chart in Fig. 5.5(d) shows how the change becomes more pronounced as the volume increases. A change in inlet pressure will affect the operation somewhat as shown in Fig. 5.5(e).

The dial of the valve shown in Fig. 5.5(a) is calibrated for easy and repeatable flow settings. Adjustments over the complete valve capacity are obtainable within a 270° arc. A dial key lock (Fig. 5.6) prevents tampering with the valve setting. The valve of Fig. 5.5(a) employs a sharp-edged orifice design making the flow immune to temperature or fluid viscosity changes.

A choice of flow ranges is usually provided to allow for maximum sensitivity within the required flow parameters. The symbol of Fig. 5.5(a) shows that models are available with a check valve for return flow if needed. Modifications may be encountered in pressure compensated flow control valves which are provided to permit use in unusual circuit applications.

A stroke limit device may be provided at either end of the reducing spool in the valve of Fig. 5.4. When applied at the end where the bias spring is located, it will limit minimum flow through the valve for rapid response and eliminate cutoff on rapid operation. An adjustment at the opposite end of the spool controls the maximum opening of the reducing spool. This limitation of travel of the reducing spool will increase the valve sensitivity. Since the valve is normally open, it must travel a certain distance from the rest position before restricting the flow to the orifice. By limiting the travel, it is possible to adjust the opening to a point at which it will pass the maximum desired flow yet reduce the response time to an absolute minimum.

The adjustment shown in Fig. 5.7, specifically designed for the valve of Fig. 5.5, is referred to as an "antilunge assembly". It can be installed in the field after the valve has been put in service if needed. Often the need is evident at the start up of the machine or when changing tooling.

Threaded Cartridges Flow regulator cartridges (Fig. 5.8) are designed primarily for subplate or sandwich valve mounting in circuits where accurate nonadjustable controlled flow is required. These cartridges are factory preset to customer flow specifications. Some models, however, include a "tuning" adjustment which allows final flow setting on the job. Typical tuning range is ±30% from the factory setting. Cartridges can be structured with or without a free-flow return check.

Relief Valve Compensation

Figure 5.9(a) shows by use of ANSI symbols how a relief valve can be the compensating agent for a meter-in flow mechanism. Figure 5.9(b) shows the full ISO symbol. Figure 5.9(c) shows the simplified ISO symbol for the flow control.

Upon reaching an unusual pressure, the relief valve in Fig. 5.9 serves a dual purpose. It will provide the safety function as an auxiliary to the compensating function. Note that the entire orifice and compound relief mechanism are contained in a single envelope. The pump delivery is directed to the adjustable orifice. An internal tee connects the input of the main spool of the relief valve to this area. The balancing line to the control chamber of the relief valve is connected from a point beyond the adjustable orifice. This pilot sensing line is orificed to prevent an undue volume of fluid from entering the small integral pilot safety valve on the relief-valve control chamber. The location of the pilot sensing line at the point beyond the adjustable orifice means that the only pressure that can be developed by the main spool of the relief valve, until flow is established downstream, will be that corresponding to the spool spring setting. The setting of the spring determines the pressure drop across the externally adjustable orifice. The downstream load is reflected into the control chamber as an additive to the main-spool spring, so that

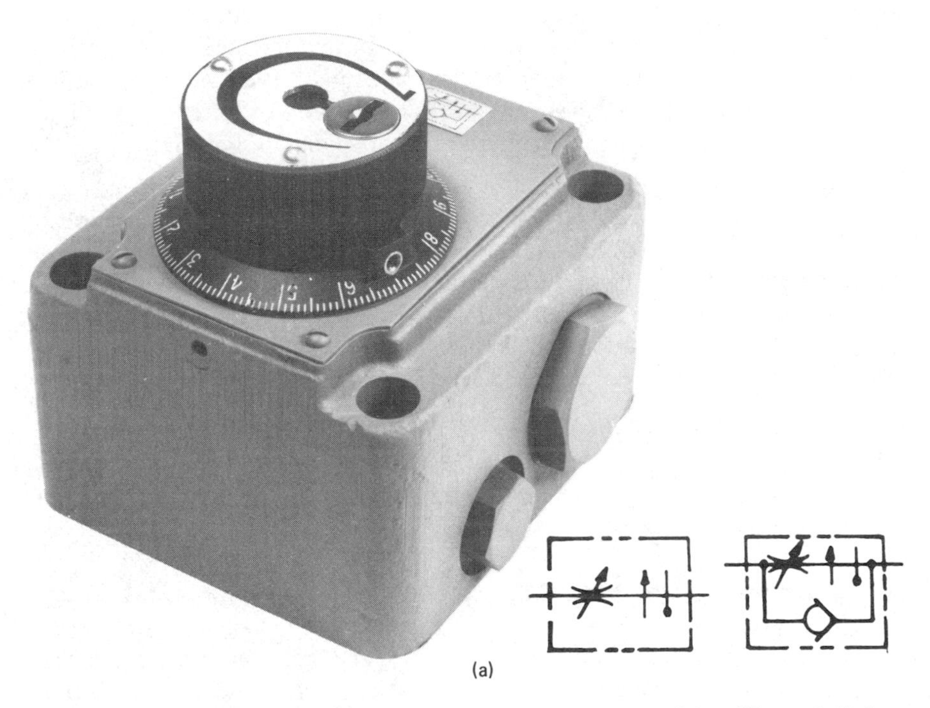

(a)

Figure 5.5 (a) Subplate mounted flow control valve with calibrated dial assembly. (Courtesy of Continental Hydraulics, Savage, Minnesota.)

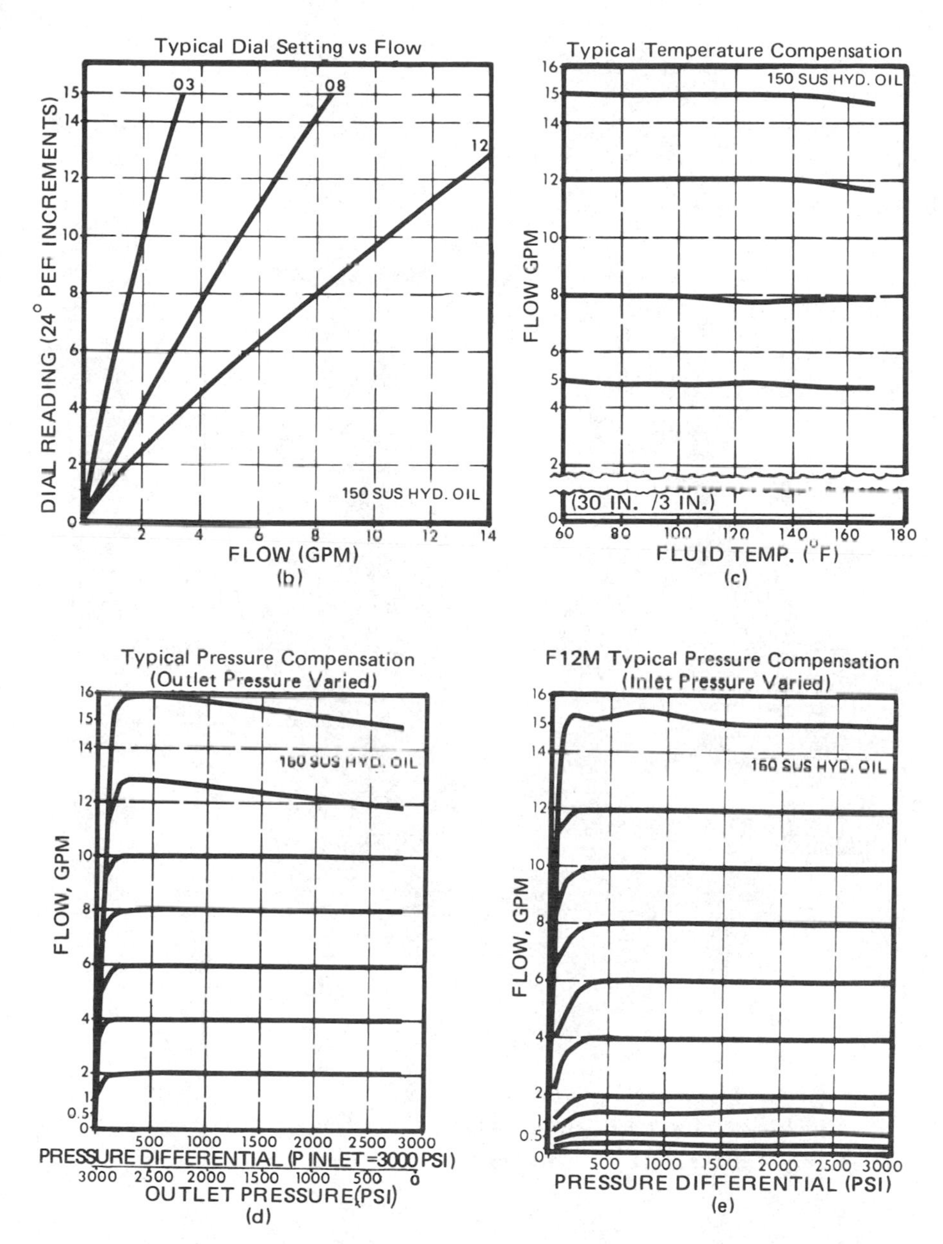

Figure 5.5 *(continued)* (b) Typical dial setting vs. flow. (c) Temperature compensation for typical pressure compensated flow control valve. (d) Typical pressure compensation with outlet pressure varied. (e) Typical pressure compensation with inlet pressure varied. (Courtesy of Continental Hydraulics, Savage, Minnesota.)

Figure 5.6 Lock prevents tampering with valve setting. Turning socket wrench clockwise in screw head locks setting of valve. (Courtesy of Continental Hydraulics, Savage, Minnesota.)

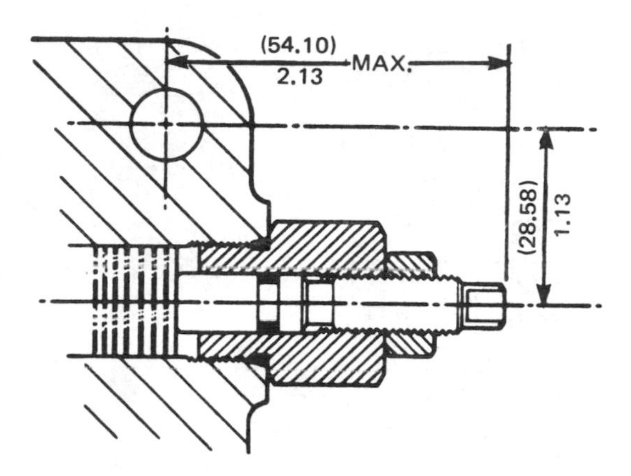

Figure 5.7 Compensator spool adjustment for maximum sensitivity. (Courtesy of Continental Hydraulics, Savage, Minnesota.)

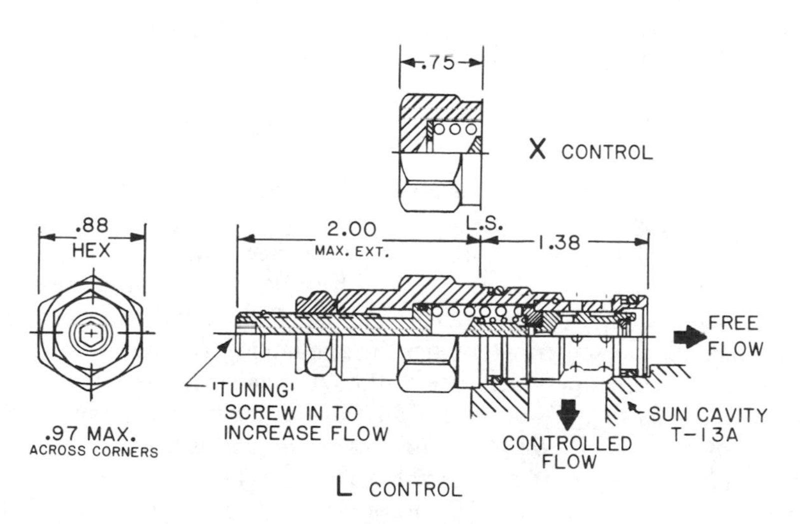

Figure 5.8 Cartridge-type pressure compensated flow regulators. (Courtesy of Sun Hydraulics Corporation, Sarasota, Florida.)

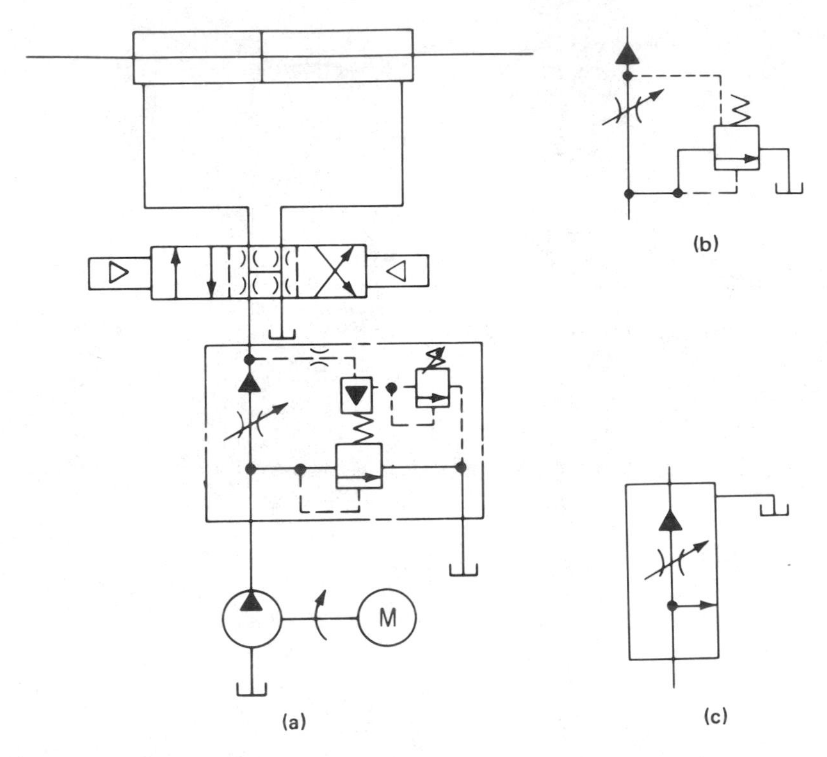

Figure 5.9 (a) Pressure-compensated flow-rate control with integral relief valve (ANSI symbols). (b) ISO symbols for flow control portion, full structure. (c) ISO simplified symbol for flow control.

the pressure drop across the adjustable orifice remains constant at the value of the spring, regardless of the downstream loading, until the sensing-line pressure becomes sufficiently high to crack through the safety relief-valve pilot; then the small, integral safety relief valve determines the maximum pressure that can be built up in the control chamber. The maximum pressure built up in the control chamber determines the maximum hydraulic additive and the maximum system pressure. Use of the relief valve and the downstream signal precludes using this device for anything but meter-in operation. Further, it can be used with only one functional unit and is usually considered as a complete circuit. It is not possible to connect two of these controls in parallel because the fluid will seek the course of least resistance and only one of the two controls can be effective.

Priority Valve Compensation

Power steering is considered essential to many off-the-road type machines. The pump supplying pressure for the power steering system may also serve other actuators on the machine. A priority flow control system insures fluid to the power steering system before it is directed to the other subsystems.

A priority circuit plus two adjustable orifices in a single housing provides an answer to several circuit needs in agricultural and road construction machinery. A step-by-step description follows for a typical system. The valving can be modified for specific problem solving.

Figure 5.10 shows the symbol for a priority valve. The spring holds the valve spool in the position shown in the left flow block. Pressure is connected to circuit 1. Circuit 1 is also connected to the control area opposing the spring. As the pressure increases in circuit 1, the control signal tends to move the spool toward the spring and divert some of the fluid to circuit 2. This is the first portion of the priority circuit. Note that the entire flow will go to circuit 2 if circuit 1 is completely blocked.

Figure 5.11 shows how the lever assembly controls an orifice metering fluid into the load circuit associated with the spring cavity on the left-hand block of the priority valve. The lever also opens a pilot two-way valve to the tank when the flow is completely stopped to circuit 1. In the position shown, the fluid signal to the spring chamber of the priority valve is vented to the tank so that the fluid will pass to circuit 2 at the pressure setting of the priority-valve spring. This can be relatively low pressure, usually about 50 psi.

As the handle is moved to meter fluid into circuit 1, a signal is reflected back to the spring chamber of the priority valve. This signal (pressure) will be the value of the load in circuit 1. The input pressure will be at least as much

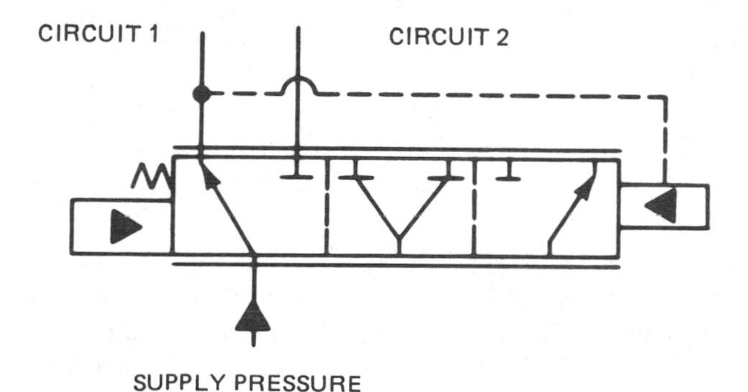

Figure 5.10 Basic compensator for priority-type flow control system.

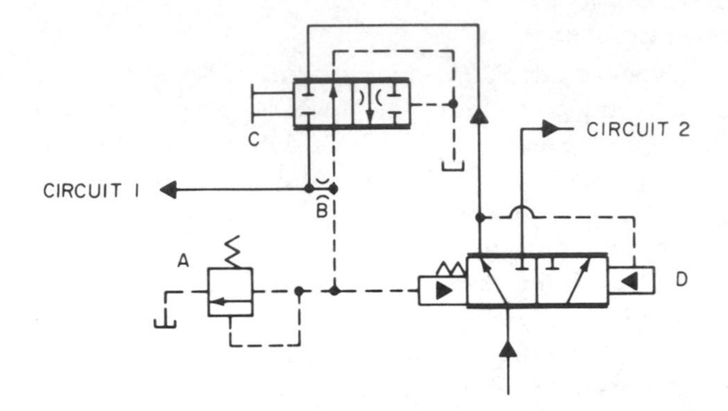

Figure 5.11 Priority valve and circuit 1.

as the spring setting plus the load pressure in circuit 1. The spring value will control the pressure drop across the orifice. A secondary orifice in the control circuit limits the amount of reflected signal fluid to a quantity that will not overload the small signal lines or pass an excessive amount of fluid through the pilot overload relief valve connected to the spring chamber of the priority valve. This small, integral, maximum-pressure, pilot-relief valve prevents excessive line loads and subsequently controls maximum pressures in this area. It is adjusted to the maximum expected load value commensurate with the circuit components and the mechanical devices being operated. This valve usually serves as a safety relief valve rather than as a maximum-pressure valve. Maximum working pressures generally are a function of loads rather than of this valve setting.

In the circuit of Fig. 5.11 the priority valve serves both as a reducing valve controlling the pressure in circuit 1 and as a diversion valve passing the remaining fluid downstream to circuit 2. It bypasses fluid and maintains the correct flow to actuate circuit 1. Further, this can reduce pressure when necessary on circuit 1 and divert fluid to circuit 2 at full pressure when needed.

Figure 5.12 omits the controls in circuit 1 to avoid complexity. The priority valve is shown to illustrate how it connects the two circuits. The valve shown immediately above the priority valve is a pilot-operated bypass valve or a relief valve. If the handle lever in circuit 2 is actuated so that the flow through the orifice is completely blocked, the lever also actuates a portion that diverts the control fluid to the spring chamber of the relief valve back to the tank. In this position any fluid diverted from circuit 1 would be passed on the tank at the value of the main-spool spring in the upper relief valve. The spring in the priority valve and the spring in the relief valve are in series so that the pressure drops are additive. However, the total value is relatively low, and little power is lost when both levers are in the closed position. The pump can be considered in the relaxed position. As the control lever of circuit 2 is moved toward the open position, the vent line or line from the control chamber of the relief valve is blocked and the reflected signal from circuit 2 load is applied to the spring chamber along with the initial spring value. This permits the excess fluid passing back to the tank at the circuit 2 load value. The safety pilot valve merely establishes a certain maximum pressure value that can be contained within circuit 2. This pressure value is usually much in excess of the work values normally encountered. Compensation for circuit 1 is by reduced pressure to the orifice controlled by the priority valve. Compensation for circuit 2 is by the input relief valve controlling the pressure drop across the orifice.

Figure 5.13 shows the complete circuit. The fluid under pressure from the pump is directed into the priority valve. This valve directs the fluid to the input of circuit 1. As the needs of circuit 1 are met by a portion of the fluid passing into this area, the excess fluid passes to circuit 2. The orifice established by the lever of circuit 1 determines the speed of operation of the mechanism fed by this circuit. Fluid being fed to the relief valve in circuit 2 meets resistance caused by a normally-closed spool and is diverted to the orifice section controlled by the position of the lever. As the pressure across the orifice establishes a maximum flow, the excess passes through the relief valve. This function occurs because the fluid in the spring chamber of the relief valve is at a controlled maximum pressure established by the work load in circuit 2. This work load plus the basic spring value determines the pressure value at which the spool can move to create an orifice of a size to bypass the excess fluid back to tank.

Levers are provided to independently adjust the flow to circuit 1 or to

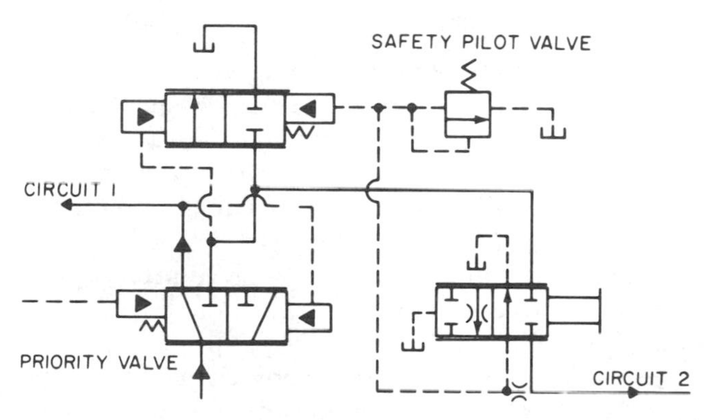

Figure 5.12 Priority valve and circuit 2.

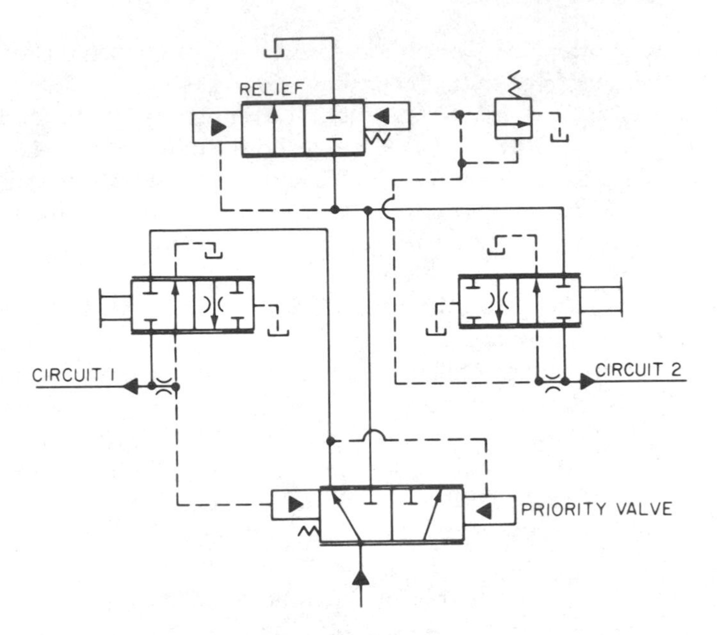

Figure 5.13 Complete priority circuit.

circuit 2. The ports consist of a pressure input, circuit 1 outlet, circuit 2 outlet, and tank outlet. The drains from the levers and safety pilot valves are connected directly to the tank port as shown in Fig. 5.13. Units of this type can be externally drained if it will make the circuit more sensitive. Often there is a maximum-movement adjustment for the levers to establish preset maximum flows and calibration for repetitive adjustments.

5.3. CONTROL ASSEMBLIES

5.3.1. Proportional Flow Control Valves

Remote and/or programmable control of an electrohydraulic flow control valve offers a high degree of design flexibility. The valve shown in Fig. 5.14 will provide an infinite variation of flow through the valve by varying the magnitude of an electrical input signal to the linear force motor.

A small, low pressure, constant flow is developed by the pilot stage of the valve shown in Fig. 5.15(a). Pilot pressure is controlled by a variable orifice (linear force motor) vented to tank and acts on the end of the main spool opposite the counterforce spring.

A direct relationship is established between the energy input impressed on the coil of the linear force motor and the flow control element.

The end result is a low cost, self-contained valve that modulates flow over a wide range without external feedback devices.

The typical specifications for a valve such as that shown in Fig. 5.15(a) are shown in Fig. 5.15(b). With this type of data it is possible to predict machine response and develop associated electrical control circuitry.

Similar valve assemblies are available using a DC motor and gear train to operate the spigot of the valve shown in Fig. 5.5(a). A potentiometer can be geared to the actuator assembly to provide remote indication of the spigot position and/or act as a part of a servoloop to remotely control flow through the valve structure.

Note the simplified symbol as shown in Fig. 5.15(c). The detailed illustration shows the relationship of the control spool to limit the pressure drop

Figure 5.14 Electrohydraulic proportional flow control valve. (Courtesy of Continental Hydraulics, Savage, Minnesota.)

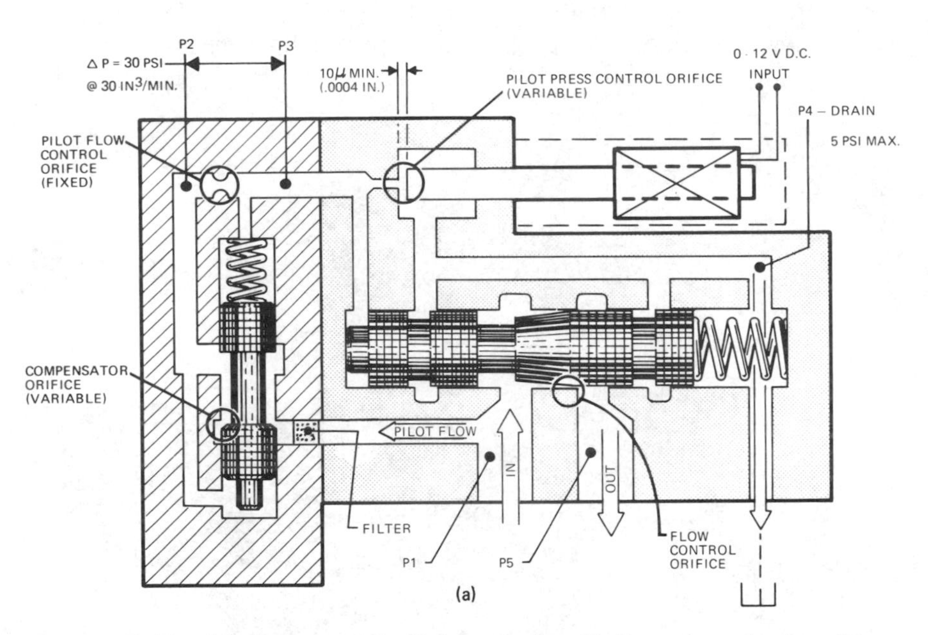

Figure 5.15 (a) Cross section of electrohydraulic flow control valve. (Courtesy of Continental Hydraulics, Savage, Minnesota.)

Voltage (nom.)	0-12 VDC	Threshold Shift (ma)	
Voltage (max.)	15 VDC	temperature change	80 - 140° F 8% of rated output
Current (nom.)		pressure change	500 - 3000 psi 4% of rated output
adjustable range	20 - 170 ma	Response Times (ms)	
Current (max.).	215 ma	0 - 100% of rated current	250 ms
Coil Resistance	70 ohms	100 - 0% of rated current	30 ms
Flow Gain (see curves below)		Repeatability	1% of rated output
Code 10	22 cipm/ma at $\Delta\rho = 75$	Linearity	3% of rated output
Code 20	43 cipm/ma at $\Delta\rho = 75$	Resolution	2% of rated output
Hysteresis (ma)	3.5% of rated output	Pilot Flow	60 cipm
Hysteresis (psi)	6.5% of rated output	*(NOTE: pilot flow remains constant at all operating conditions)*	
Threshold (ma)	20% of rated output		

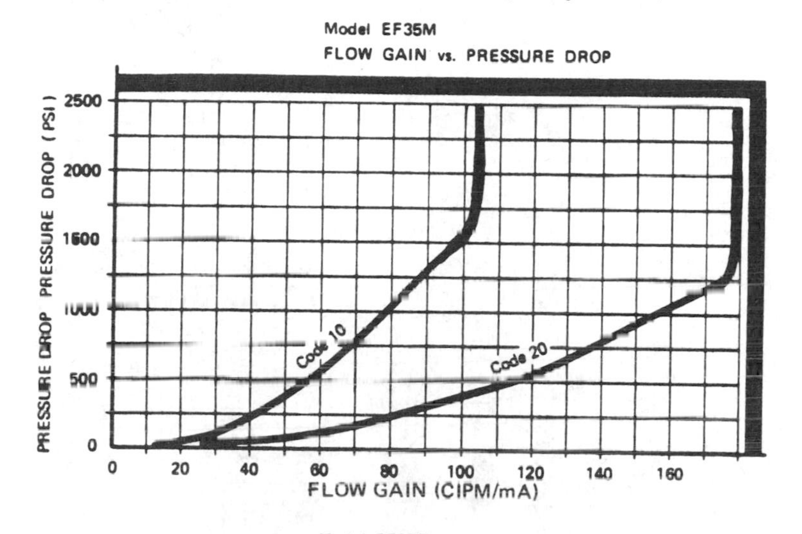

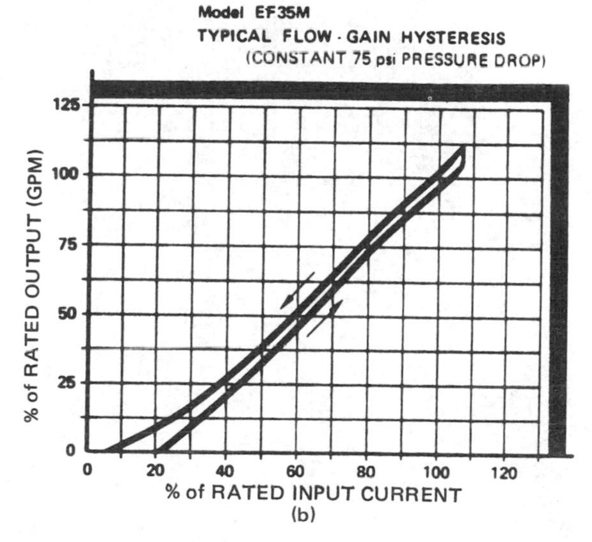

Figure 5.15 *(continued)* (b) Technical data and response curves for the electrohydraulic flow control valve of Fig. 5.15(a). (Courtesy of Continental Hydraulics, Savage, Minnesota.)

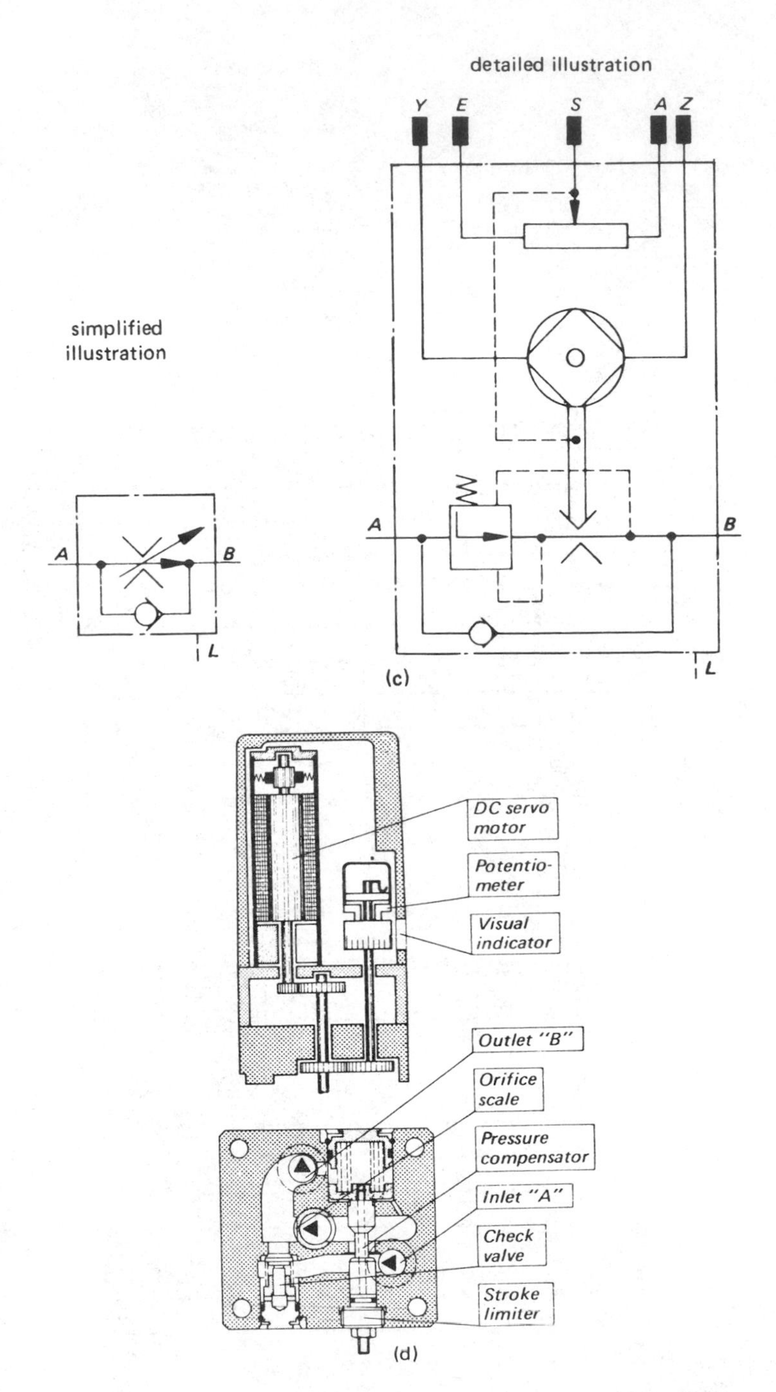

Figure 5.15 *(continued)* (c) Direct current motor actuation of flow control valve. (d) Cross section view of valve shown in symbology in Fig. 5.15(c). (Courtesy of The Rexroth Corporation, Bethlehem, Pennsylvania.)

across the orifice which is adjusted by the DC motor assembly. Whichever point of control is appropriate to the machine function can provide the signal to the DC motor. A visual indicator at the valve provides on sight information for the service personnel. The potentiometer is shown as a geared mechanism in direct relationship with the visual indicator providing information as to the spigot position.

The cross section drawing of 5.15(d) is quite similar to that of Fig. 5.4. The integral check valve assembly is shown in Fig. 5.15(d).

5.3.2. Rapid Traverse and Feed

Many combinations of directional and flow control valves are integrated into a single envelope or built into a machine control panel to provide rapid traverse of a machine slide, one or more feed rates, a dwell, and rapid return. These units are used for milling, drilling, reaming, boring, and other similar machining operations.

The units may employ one compensating mechanism and a device to divert flow from one orifice to two or more to provide desired machine feed rates. The directional control valve may be controlled manually, by cams, solenoids, or a combination thereof. The basic components integrated into the circuit follow the design parameters presented in the preceding chapters.

5.3.3. Reciprocating Controls

Grinding machines and other devices needing a continuously reciprocating motion pattern often use a circuit similar to that of Fig. 5.9. The directional control and the relief valve compensated flow control may be integrated into a single envelope. Directional control may be by a cam actuated device for precise reversal. Venting of the relief valve can be employed to stop and start the machine function.

5.3.4. Acceleration and Deceleration Control

Flow control devices can become an important part of directional control valve assemblies. Flow control modules are integrated into the valve assembly of Fig. 5.16. These flow control modules are integrated into auxiliary pilot control piston assemblies. This offers extremly flexible control of the main spool. The double solenoid actuated, spring-centered four-way hydraulic pilot valve controls the direction of movement of the major four-way directional control spool.

Integrated into the valve assembly of Fig. 5.16 is a basic fluid-piloted spool type four-way valve. The main spool is spring centered. The auxiliary pilot piston assembly is built into the two end caps as shown in Fig. 5.17. The auxiliary pilot piston is supplied with a mechanical adjustable screw-type stop in each direction of movement.

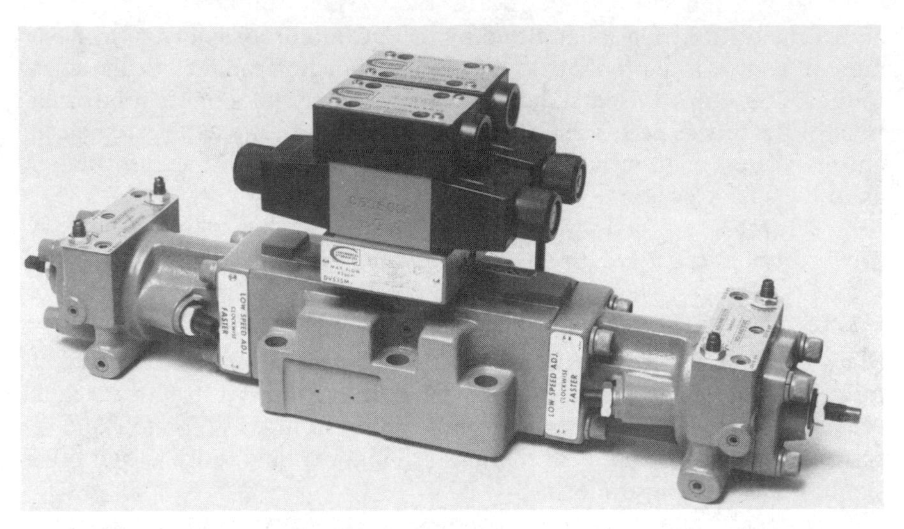

Figure 5.16 Electrohydraulic directional control valve with multiple acceleration and deceleration controls. (Courtesy of Continental Hydraulics, Savage, Minnesota.)

The direct solenoid operated three-position spring-centered pilot valve is manifolded to the main directional valve. Pilot lines from the three-position pilot valve are interconnected to the spring pockets at the end of the main spool. The main directional spool will respond to a pilot (pressure) signal from the three-position pilot valve.

The distance of travel of the main spool responding to the pressure signal from the solenoid actuated pilot valve will be subject to the position of the auxiliary pilot pistons in the end caps.

A single solenoid four-way pilot valve, used in this assembly as a three-way valve, is also integrated into the assembly of Fig. 5.16. This single solenoid pilot valve has but two functions. One function is to direct pressurized fluid to both pilot end caps simultaneously. The second function is to release pressurized fluid from both pilot end caps simultaneously.

The pilot end caps each contain two flow control valves and two check valves plus the mechanical stops in each direction of travel.

The effective diameter of the pilot piston in each end cap is greater than the diameter of the main spool. Figure 5.17(a) shows the conditions within the valve structure of Fig. 5.16 when pressure is applied at the port supplied by the pump and all solenoids are deenergized. The fluid lines to the actuator are locked in position by the closed passage at the main-spool assembly. Both pilot pistons are pressurized toward the main spool but are restrained by the slow speed stop adjustment, and do not touch the spring-centered spool.

Solenoid A has been energized in Fig. 5.17(b). The main spool shifts. It is restrained by the large diameter pilot piston in the end cap at a *creep flow rate* resulting from the position of the slow speed adjustment. The actuator will move at this creep rate until Solenoid B_1 is energized as shown in Fig. 5.17(c).

By energizing solenoid B_1, the restraint of the pilot piston in the end cap is released at a controlled rate through the integral acceleration adjustment. The main spool will continue to increase major directional flow at a controlled acceleration rate until the adjustable full speed stop is reached. A controlled maximum flow is established to and from the actuator by the main directional spool. To decelerate movement rate, solenoid B_1 is deenergized [Fig. 5.17(b)]. Pressure at a controlled flow rate enters the large diameter pilot piston forcing the main spool back to the creep position. At this position solenoid A is deenergized, the main-spool spring centers, and work load stops. The same pattern is used for reverse flow action.

Energizing solenoid B reverses direction of main-spool travel and the same pattern of acceleration and deceleration occurs on the other end cap. The rate is established independently at the pilot piston assembly on the other end cap.

5.3.5. Flow Sensitive Valves

Flow fuses are designed to protect a hydraulic system from loss of fluid if a line breaks. The valve stays open and allows flow between the system and an actuator at normal flows, but closes and shuts off all flow instantly if a line breaks and flow from the actuator exceeds the setting of the valve; this will lock the actuator in place and protect personnel in the area of the break.

Flow fuse cartridges may be installed directly into cavities machined in the cylinder assembly, line mounted, or included as part of a circuit in a manifold. Figure 5.18(a) shows the construction of a typical flow fuse. Figure 5.18(b) shows the identifying symbol.

5.4. SUMMARY

Flow control mechanisms are used to control the rate of movement of an actuator. Flow control mechanisms can be integrated into directional control valve assemblies. Directional control valves can also control rate of actuator movement.

Various combinations of flow and directional control valves can be structured to provide precise control of acceleration, deceleration, and movement patterns of actuators. Adjustments are usually infinitely variable from rest to maximum movement rate.

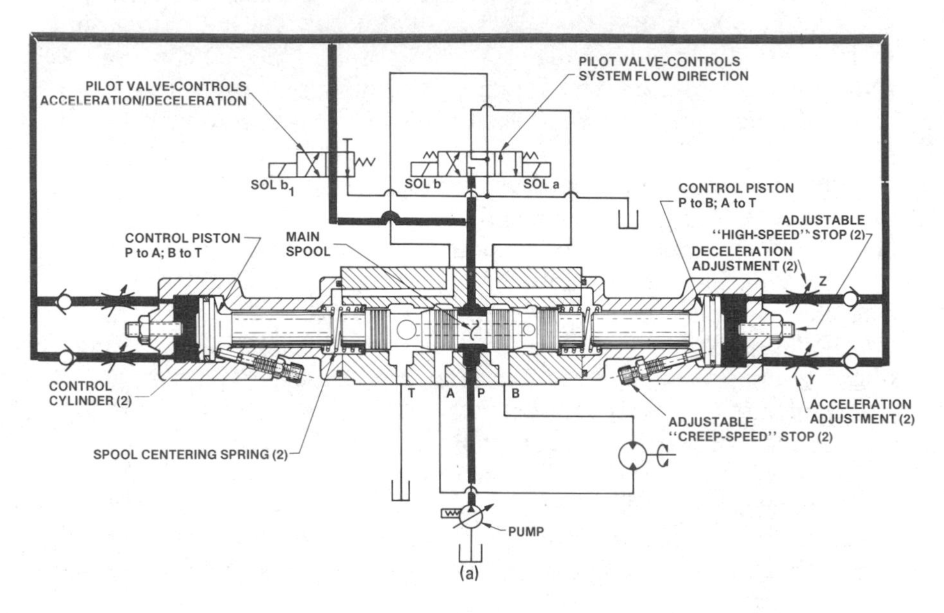

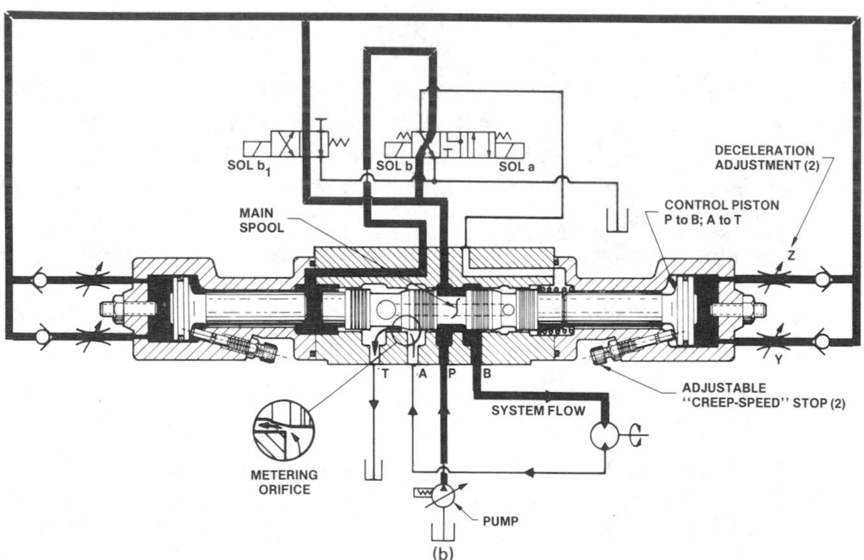

Figure 5.17 (a) Valve of Fig. 5.16 in neutral or stop position. (b) Valve of Fig. 5.16 with main spool shifted to creep speed position.

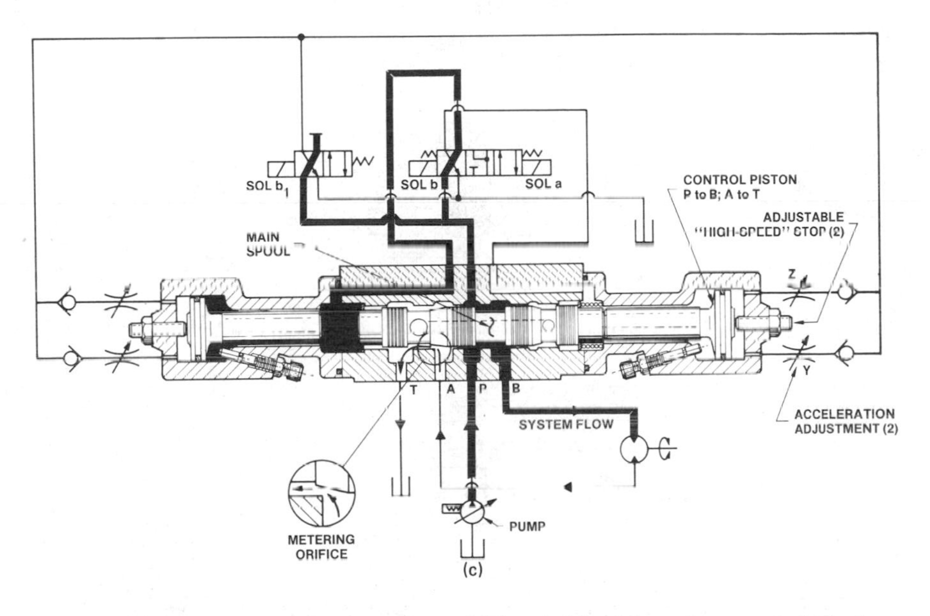

Figure 5.17 *(continued)* (c) Valve of Fig. 5.16 with main spool shifted to full speed position. (Courtesy of Continental Hydraulics, Savage, Minnesota.)

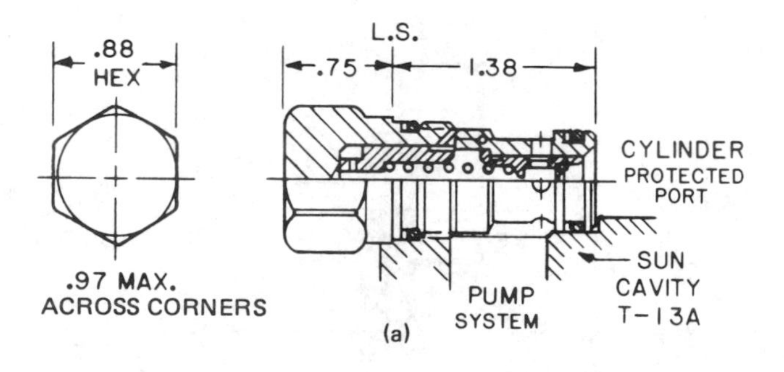

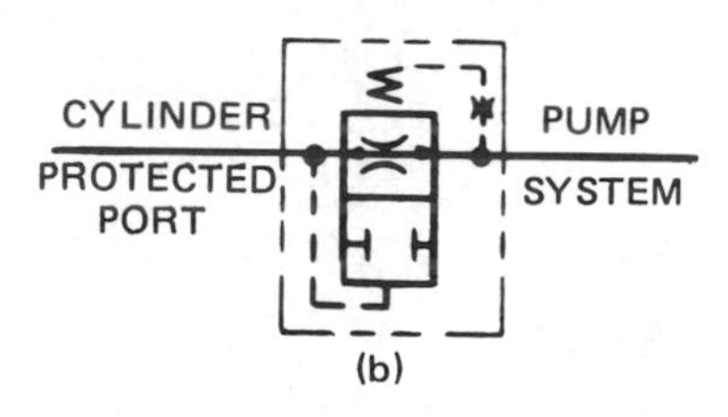

Figure 5.18 (a) Flow fuse cartridge. (b) Symbol for flow fuse. (Courtesy of Sun Hydraulics Corporation, Sarasota, Florida.)

6 Servo Control Systems

6.1 INTRODUCTION

6.1.1. Characteristics of a Servo System

What is a servo? In its simplest form, a servo (more properly, a *servomechanism*) is a control system which measures its own output and forces the output to quickly and accurately follow a command signal. In this way, the effects of anomalies in the control device itself and in the load can be minimized, as can the influence of external disturbances. Servos can be designed to control almost any physical quantity (e.g., motion, force, pressure, temperature, electrical voltage, or current).

This chapter is concerned with servos designed to control mechanical loads.

When rapid, precise control of sizable loads is required, an electrohydraulic servo is often the best approach to the problem. The basic elements of such a servo are shown in Fig. 6.1(a).

The output of the servo is measured with a transducing device to convert it to an electrical signal. This feedback signal is compared with the command signal, and the resulting error signal is then amplified and used to drive the servovalve. The servovalve controls the oil flow to the actuator in proportion to the drive current from the amplifier. The actuator then forces the load to move. Thus a change in the command signal generates an error signal which causes the load to move in an attempt to zero the error signal. If the amplifier gain is high, the output will very rapidly and accurately follow the command, even in the presence of such annoyances as servovalve null shift and load friction.

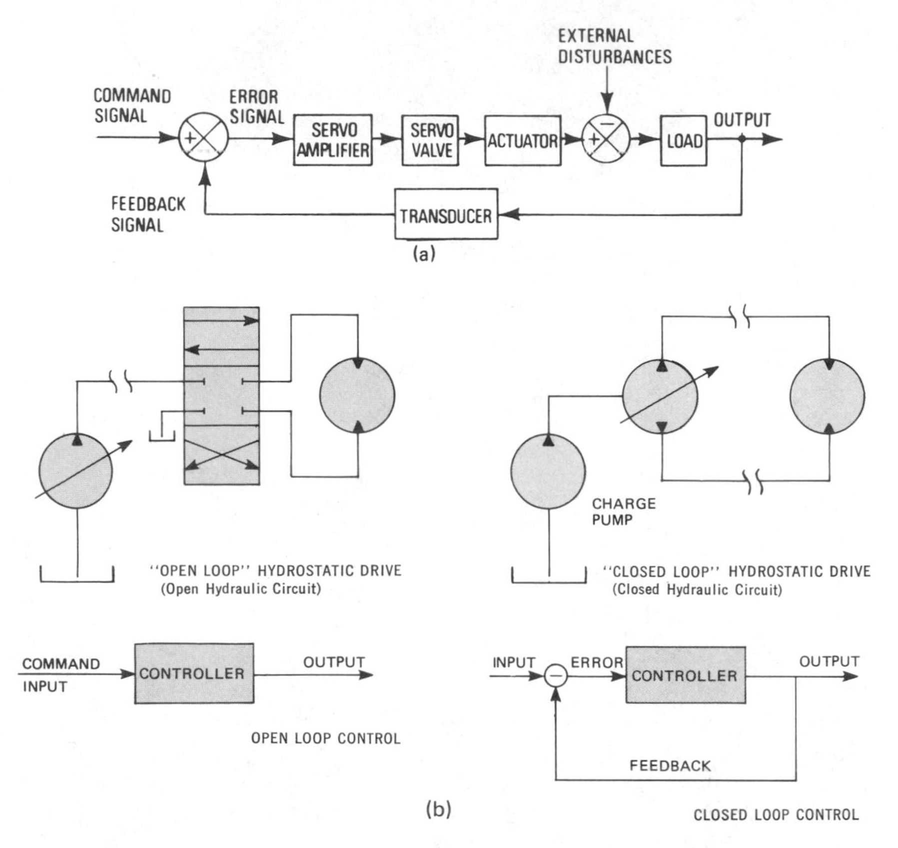

Figure 6.1 (a) Basic electrohydraulic servomechanism. (b) Open loop vs closed loop. (Courtesy of Moog Inc., East Aurora, New York).

External disturbances can cause the load to move without any change in the command signal. To offset the disturbance input, an actuator output is needed in the opposite direction [*See* Fig. 6.1(a)]. To provide this opposing output, a finite error signal is required. The magnitude of the required error signal is minimized if the amplifier gain is high.

Ideally, the amplifier gain would be set high enough that the accuracy of the servo becomes dependent only upon the accuracy of the transducer itself. In practice, however, the amplifier gain is limited by stability considerations. In some applications, stability may be critical enough that the desired performance is not possible, even with closed loop control. That is, closed loop control is not a panacea. Performance estimates are given in the following sections for the three most common types of electrohydraulic servos: position (linear or angular), velocity (linear or angular), and force (or torque).

Estimates are first made for the linear cases, and conversions are later given for the corresponding angular quantities.

We recognize that all power transmission systems must have known control characteristics. Fluid power transmission systems can respond to a multiplicity of controls covering very wide parameters.

Servo control systems are thus used to provide a master-slave relationship. A tiny signal dictating rate of movement, direction of movement, and force level can be amplified to many times the initial signal value through the use of this type of fluid power system.

The servo-type fluid power systems considered in this chapter can respond to a manual input signal such as power steering used for directional control. Power brakes offer a typical example of force amplification. Speed control devices manually set by a potentiometer adjustment can be kept in synchronization by tachometer feedback in the motion output area.

Sensitivity of response to an input signal is governed by component characteristics and associated economic considerations.

A directional control device, as an example, for economic and other reasons, may have significant seal in the neutral position. The linear movement of a spool-type valve required to uncover the valved ports is often referred to as the *dead band*. Servovalves are typically structured with line-to-line contact or possibly a *negative* lap to increase sensitivity. Fluid losses through leakage are usually increased as sensitivity is gained by the reduced dead band.

Systems with mirror-image-type response from many different input signals will undoubtedly include predetermined control losses as a cost of the needed control sensitivity. As an example, tracer-type devices can follow templates to guide work mechanisms such as flame cutters used to develop shapes from steel plate, milling cutters, and grinders used in the production of various types of dies and similar manufacturing operations.

Military uses for aiming guns provided the basis for control of missiles and aircraft. These applications fostered innumerable industrial applications needing similar speed and accuracy.

Thus a fluid power servo system can faithfully reproduce any input signal repetitively at any degree of accuracy commensurate with the basic system capabilities. The input can be manual, a preprogrammed tape source, or the system can respond to guidance from a surveyor line, laser beam, or any signal that can be identified, interfaced, and accepted by the system.

6.1.2. Open Loop versus Closed Loop Systems

Terminology widely used in certain areas where hydraulic power transmission has historically been associated with manual controls conflicts to some extent with the circuitry associated with servo systems and especially those responding to a feedback device.

The published glossary of terms for fluid power [1] does not define *open loop* or *closed loop* in the section covering circuits, diagrams, systems, standards, and codes.

W. J. Thayer, in a paper presented at the Colorado School of Mines [2], offered some observations which help to clarify the use of these terms. He referred to the fact that conventional hydraulic circuits generally use a relatively large reservoir. The pump is supplied with fluid from this reservoir with gravity flow and/or atmospheric pressure pushing the fluid into the intake line because of the pumping action. Return fluid from the actuator is directed to the reservoir. Generally a baffle is provided between the return flow and the pump suction line. This type of circuit is often referred to as an *open loop* hydraulic circuit. For clarification the term *open hydraulic circuit* can be used with the term open loop reserved to describe an *open loop control.*

Figure 6.1(b) illustrates the "open loop" hydrostatic drive which might better be referred to as an open hydraulic circuit. The lower block representation illustrates the generally accepted *open loop control.* Likewise, the adjacent hydrostatic transmission which directs return flow from the actuator (hydraulic motor) to the intake of the pump is shown in the same figure as a closed hydraulic circuit rather than "closed loop" hydrostatic drive. The lower figure introduces through the block diagram the *feedback function* indicating the *closed loop control.*

As a further clarification, Fig. 6.2 shows a closed hydraulic circuit with facilities for an open loop hydrostatic control consisting of a power source, control station, and pump controller.

Figure 6.3 shows the closed hydraulic circuit of Fig. 6.2 equipped with a tachometer feedback signal relating to a command setting with the error signal being fed to the pump controller to close the loop and establish a speed in accordance with the command setting. Thus there is a closed loop hydrostatic control.

In summary: The simple connection of a potentiometer and battery to a hydraulic pump containing an electric controller gives remote velocity control for a hydraulic motor (Fig. 6.2). Control is open loop. If the engine slows down, or if the load builds up, then the load velocity will change. Of course the operator may "close the loop" by observing the output velocity and moving the control lever to compensate, but this is not automatic feedback control.

If a tachometer is attached to measure the speed of the hydraulic motor, then the tachometer signal can be fed back and compared with (subtracted from) the command signal (Fig. 6.3).The resultant error signal is amplified and used to drive the pump controller. Best performance will be achieved if

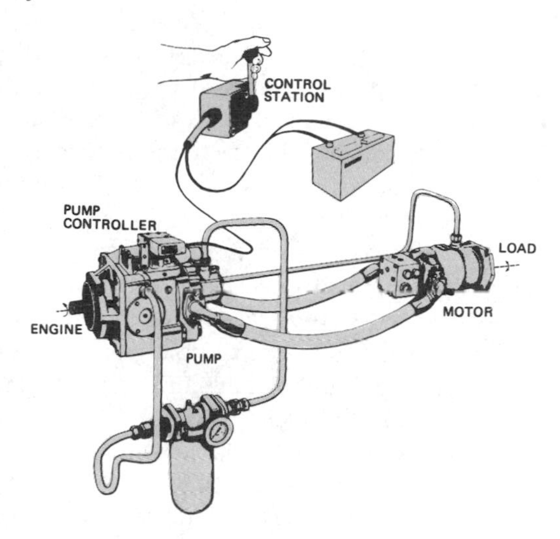

Figure 6.2 Open loop hydrostatic control. (Courtesy of Moog Inc., East Aurora, New York).

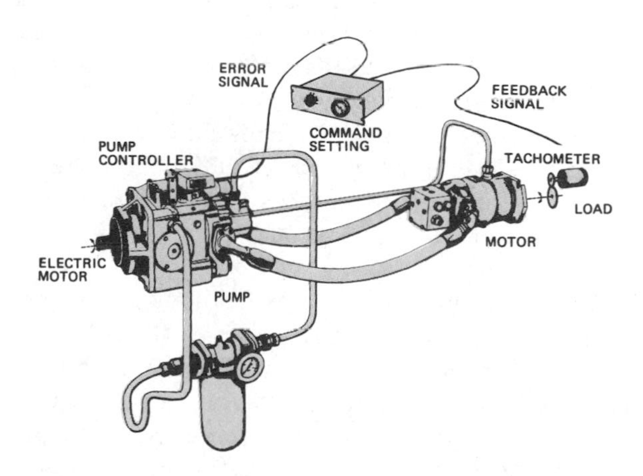

Figure 6.3 Closed loop hydrostatic control. (Courtesy of Moog Inc., East Aurora, New York).

the amplifier has integral control. Thus the amplifier supplies a continuously increasing amount of current to the controller whenever a steady velocity error exists. This system is called a velocity servo. The other common type of servo is the position servo. In a position servo the electrical command from a potentiometer, or other source, sets the desired position of the load driven by a hydraulic motor (or cylinder and piston if preferred).

Another potentiometer or other position measuring transducer (like a linear variable differential transformer LVDT or DCDT which is a linear variable differential transformer with integral solid-state exciter and phase sensitive demodulator) is attached to measure the load position. Again, this output signal is fed back, compared to the command signal, and the difference signal is amplified to drive the controller.

The primary advantage of closed loop control is improved accuracy. Any drifts in the controller, or leakage in the hydraulic pump or motor, or loading down of the drive, are automatically detected and the controller is adjusted to offset the change.

T. P. Neal, in a paper presented at the National Conference on Fluid Power [3], has clarified some areas relative to performance estimation for electrohydraulic control systems. Some of the calculations in this chapter and examples are from Neal's presentation. Neal makes the observation that when closed loop electrohydraulic control systems first began to appear in industry, the applications were generally those in which very high performance was required. While electrohydraulic servos are still heavily used in high-performance applications such as machine tool industry, they are beginning to gain wide acceptance in a variety of industries. Examples are plastics, oil exploration, mining, automotive testing, materials handling, and mobile equipment. In some cases the customer has need for performance which simply cannot be achieved with open loop hydraulic controls or with other types of closed loop controls. In other cases, complicated hydraulic circuitry can be replaced by electronic logic, resulting in improved reliability and lower cost. In any case it is often difficult for the potential user to assess what performance he can expect from closed loop electrohydraulic controls, or indeed whether he should even consider using them.

The purpose of this chapter is to provide the potential user with the means to make rough performance estimates for the most basic types of servos. Of course, there are endless variations of the basic types, and detailed performance prediction can often be a complex business. Because of this, the estimation techniques offered should not be considered a substitute for the services of a competent engineer with a background in servo devices and systems.

6.2. CIRCUIT TYPES

6.2.1. Open Loop

Manual Signal

The act of pushing or pulling the manual lever of Fig. 6.4(a) initiates a movement which changes the position of a spool-type four-way directional control valve. This valve directs pressurized fluid to one of the two control pistons. These control pistons actuate and control the position of the pump swashplate which in turn determines the displacement of the pump rotating assembly. Simultaneously, the drag-link moves. The action of this linkage causes movement of the four-way valve spool toward the neutral position. A direct relationship is established between the position of the manual control lever and the pump swashplate. Significant energy is required to move the swashplate. This is provided by the pressurized fluid acting on the respective control piston. Energy needed to move the manual control lever is very modest. Any linear or limited rotation actuator can be controlled by a similar system.

Electrical Signal

The pump shown in Fig. 6.2 is equipped with an *electrohydraulic servovalve*. The operator is actuating a lever which controls electrical energy flow to the servovalve which in turn determines the position of the pump swashplate and the relative speed of the motor.

Solenoid Valves On/off valves are generally solenoid controlled, often through a poppet-type, three-way valve used as a pilot. Because of the inherent full on or closed-off nature of these valves, their use is generally restricted to open loop control. Examples include positioning booms in small utility trucks and materials loaders; operating accessories such as outriggers and winches; and switching hydraulic circuits (controlling blocking valves, bypass valves, etc.).

Solenoid valves are available for both AC and DC operation at a number of different voltages. Most solenoid valves require several watts of electrical power, and, therefore use of a relay for control.

Human operators often achieve quasi-proportional control using solenoid valves by turning the valves rapidly on and off, or by jogging the valves when trying to improve feathering or resolution. Dual solenoids on a single directional valve are necessary for forward/off/reverse control.

Proportional Valves Four-way directional valves which may have an electrically operated *pilot stage,* and give valve spool position proportional in magnitude and polarity to the electrical input, are called proportional valves. As

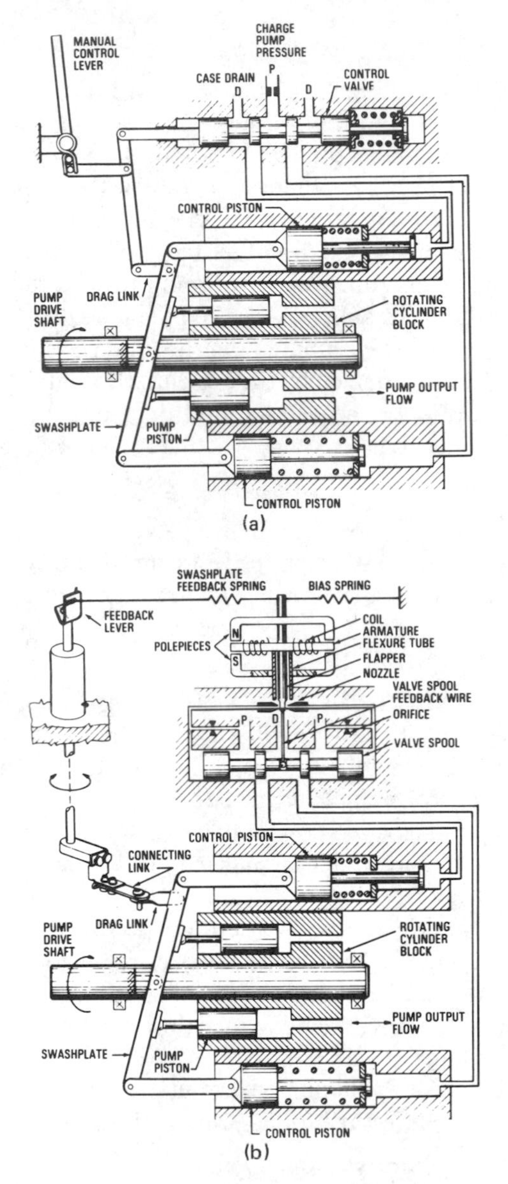

Figure 6.4 (a) Conventional hydrostatic pump controller. (b) Electrohydraulic hydrostatic pump controller. (Courtesy of Moog Inc., East Aurora, New York).

an example, the pilot section of the valve of Fig. 6.5 is fitted with direct current solenoids. The pilot spool is supplied with pressure from the major input to the four-way slave spool. The pilot spool is self centering by flow forces if neither solenoid is energized. Note the passage into the solenoid armature cavity (the armature is in the working fluid within a tubular structure. The coil is outside the nonmagnetic tubular structure). The fluid directed to the end cavities adjacent to the main slave spool in which the centering springs are contained is also directed to the end of the pilot spool and armature. The switch is actuated to direct current to the right-hand solenoid. Assuming a 28 V winding on the coil, it follows that 14 V will shift the slave spool from neutral to a half-shifted position. The pilot spool diameter sees the pressure in the conductor to the end of the slave spool. The centering spring urging the slave spool to neutral is compressed at the right end as pressurized fluid is directed to the left spring pocket. The amount that the spool moves is related to this spring and the pressure level in the spring pocket at the opposite end of the slave spool. This pressure is also effective on the end of the left end of the pilot spool which urges the spool back to a position to lock the fluid at the desired position until the current flow to the solenoid is increased or decreased with spool movement change appropriate to the current value to the solenoid. Similar action occurs with current directed to the left solenoid. As the solenoid is deenergized, the spring force on the slave spool creates a pressure which centers the pilot spool and permits the slave spool to return to the centered position. The pilot drain is returned to the major re-

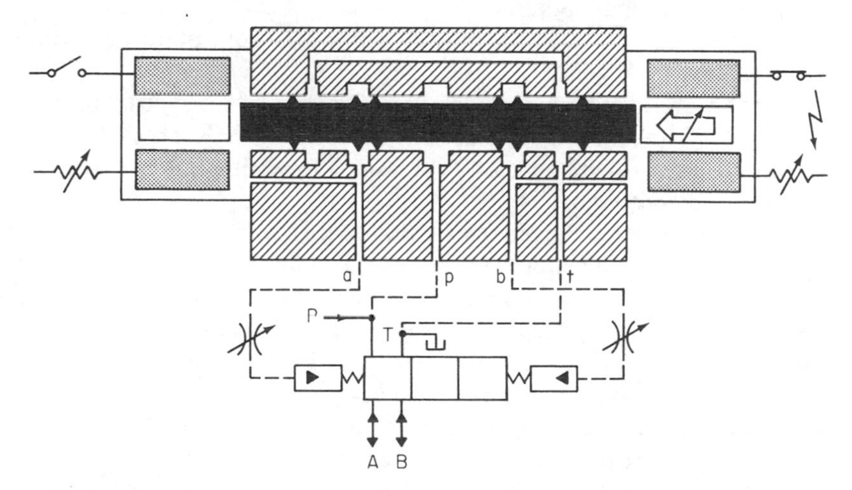

Figure 6.5 Proportional control valve. (Courtesy of HPI/Nichols, Sturtevant, Wisconsin).

turn to tank line. A separate drain can be used if excessive back pressure is anticipated in the major return to tank lines. These valves differ from digital on-off-type solenoid operated pilot valves with finite shifted and rest positions and rapid movement from position to position with minimum usable control speeds of spool movement. The type of pilot used may be a three-way structure or pilot valve with hydraulic feedback as per Fig. 6.5. The spool configuration used is different and the mechanism employed to give proportionality is usually a direct current solenoid.

In proportional valves the pilot stage may be separately pressurized by a pilot pressure supply, or it may be supplied from the normal valve supply pressure as shown in Fig. 6.5. The pilot stage requires good filtration so a separately pressurized pilot allows use of a low capacity, fine filter, rather than filtering full flow on the pressure side. Ten micron filtration is required for some pilot valves. Some manufacturers suggest 3 to 5 μm filtration for the best valve performance.

6.2.2. Closed Loop

The closed loop hydrostatic control of Fig. 6.3 can employ a pump swashplate control as illustrated in Fig. 6.4(b).

In this structure the *torque motor* (Fig. 6.18 below) is supplied with an electrical signal. The movement of the piloted directional valve directs pressurized fluid flow to the selected control piston. Movement of the swashplate actuates the draglink. A feedback lever is connected to the bias spring assembly to bring the piloted valve to neutral and hold the swashplate in a position related to the input signal.

This is a position servo. The desired action is to control the position of the swashplate.

Probably the most basic closed loop control system is the position servo. This section will discuss the various factors affecting the performance of a typical position servo, and develop expressions for estimating the more important performance items.

It is very instructive to first examine the static characteristics of the servoloop (neglect the dynamic response of the individual components in the loop). The heart of the servo system is the servovalve, and it is essential that its characteristics be thoroughly understood. Figure 6.6 shows the static characteristics of a typical critical-center servovalve. The servovalve output spool is positioned proportional to the electric current applied to the torque motor coils. Movement of the spool opens a metering orifice from the constant supply pressure P_s, to the piston; and an identical orifice connects the other side of the piston to return (tank). For no-load conditions ($\Delta P = 0$), the servovalve output flow is linearly proportional to the size of the metering orifices, which

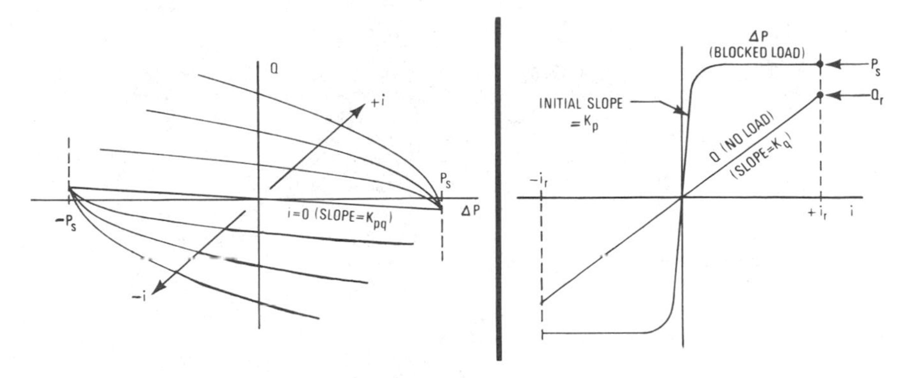

Figure 6.6 Static characteristics of a typical flow control servovalve. (Courtesy of Moog Inc., East Aurora, New York).

are proportional to input current. When a load opposes movement of the piston, the flow for a given input current is reduced somewhat, but the droop of the curves is not large until the load approaches P_s. For operation near null (i near zero), the droop of the curves approaches K_{pq}, which is determined primarily by spool null leakage (actuator leakage will add to K_{pq}). The small droop results in a very high pressure gain, K_p. In fact, the linear extension of K_p reaches P_s for input currents which typically range from 2% to 5% of rated current. This means that $K_{pq} = 0.02$ to 0.05 (Q_r/P_s). The servovalve characteristics near null are particularly important for a position servo since it spends a large percentage of the time near zero velocity (zero flow).

A schematic diagram of a complete position servo is shown in Figure 6.7. The idealized block diagram of Fig. 6.8 is developed from the schematic

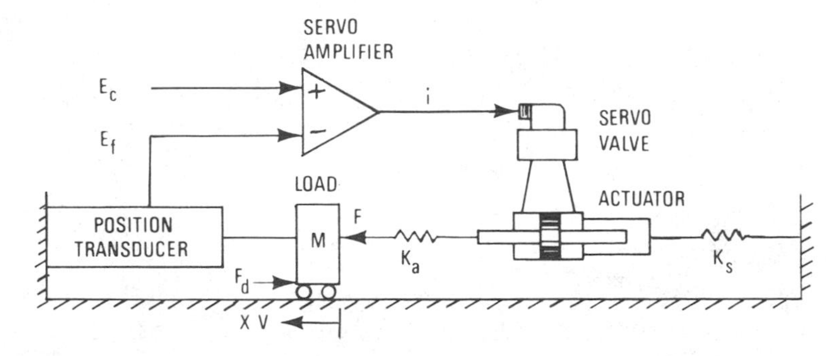

Figure 6.7 Schematic diagram of typical position servo. (Courtesy of Moog Inc., East Aurora, New York).

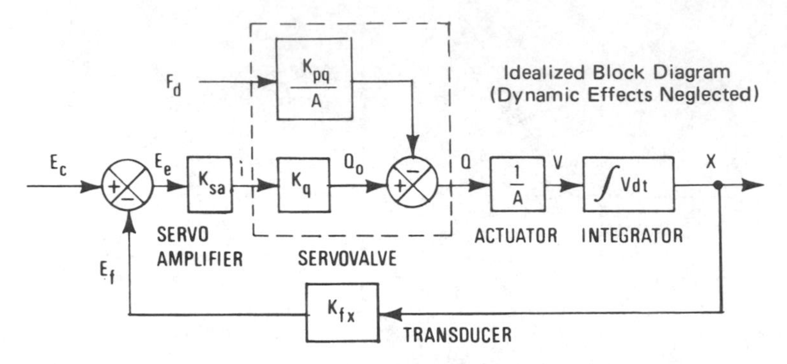

Figure 6.8 Typical position servo. (Courtesy of Moog Inc., East Aurora, New York).

by neglecting transducer dynamics, and load mass. (Note that a complete list of symbols is given at the end of the chapter). Probably the most obvious performance item is the closed loop response to command signals ($F_d = 0$). First, it is important to recognize that the loop shown in Fig. 6.8 is a zero-error type (in the absence of force disturbances) due to the integration in the forward loop. That is, even a small error signal will cause the actuator to move at a velocity. Since output position is the quantity being fed back, the feedback signal will continue to grow until the velocity becomes zero. Obviously, the velocity can only become zero when the error signal becomes zero (i.e., the load is positioned exactly where it is commanded to be). Therefore, the steady-state response of the loop is $X = (1/K_{fx})E_c$. The dynamic response of this loop is primarily dependent upon the product of the gains around the loop. Because of the integration in the forward loop, this product is called the loop velocity gain, K_{vx}, which has units of sec^{-1}. If the output position were temporarily disturbed by a transient force (with $E_c = 0$), the loop velocity gain is a direct measure of how much velocity the servo will generate in correcting the error ($V = K_{vx}X$). The loop velocity gain can be calculated as follows:

$$K_{vx} = \frac{K_{fx}K_{sa}K_q}{A} \tag{1}$$

Conventional block diagram algebra can be used to show that the closed loop dynamic response is a simple first-order lag with a time constant equal to $(1/K_{vx})$. A frequency response would exhibit 45° of phase lag at a frequency of $(K_{vx}/2\pi)$ H. It should be noted that this discussion relates to the servo response to small command signal changes. As the command magnitude increases, the response will generally become limited by the maximum servovalve flow.

The next basic performance item to be considered is closed loop stiffness (i.e., the amount of external force required to cause a position error). Again, for a steady-state condition to be achieved, the input to the integrator must become zero. Thus, the position response to an external force alone ($E_c = 0$) can be determined by setting $Q = 0$ and the output stiffness then becomes:

$$\frac{F_d}{X} = A^2 \left[\frac{K_{vx}}{K_{pq}}\right] \tag{2}$$

As mentioned previously, K_{pq} for a critical-center servovalve is generally 0.02 to 0.05 (Q_r/P_s). Using the more conservative value, the stiffness expression reduces to:

$$\frac{F_d}{X} = 20K_{vx} \left[\frac{AP_s}{Q_r/A}\right] \tag{3}$$

Thus, closed loop stiffness is proportional to the loop velocity gain and to the ratio of maximum output force/maximum output velocity.

The drive stiffness terms shown in Fig. 6.7 do not directly appear in the expressions for closed loop stiffness. The reason for this is that the position transducer shown measures load position relative to ground (i.e., the stiffness terms are inside the loop). Drive compliance inside the loop simply means that the servovalve must flow more oil when compensating for a disturbance force. However, drive compliance *indirectly* influences closed loop stiffness by reducing the maximum stable value of K_{vx}. Also it should be noted that it is not always practical to mount a transducer so that it measures load position relative to ground. For example, it may only be possible to measure load position relative to the actuator body. In this case, the stiffness calculated in Eq. (3) is in series with K_s, so that the system can never be made stiffer than K_s. Thus, location of the transducer can have a very strong influence on overall system stiffness.

Having now discussed servo load stiffness and response to command signals, the next logical topic is static accuracy. Certainly load stiffness is a factor in static accuracy, but anomalies in the servovalve and drive system can cause inaccuracies even in the absence of force disturbances. Examples of servovalve anomalies are hysteresis, null shifts with temperature and supply pressure, and threshold. Examples of actuator and drive system characteristics which can cause inaccuracies are friction and lost motion. Lost motion in the drive system (or transducer linkage) is potentially the most serious from an accuracy standpoint. This is particularly true if the lost motion is inside the servoloop, because it can easily cause low-amplitude limit cycling which severely limits the maximum usable value of K_{vx}. Stability problems due to lost motion are very difficult to predict, and every effort should be made to achieve a tight drive system. An integrated servoactuator assembly effectively eliminates most lost motion within the servoloop.

Servovalve threshold can cause problems similar to lost motion, except that a good servovalve design will hold threshold to very small values. Assuming that lost motion is held to small values, anomalies in a well-designed servovalve should total no more than 5% of rated current.Another 5% of rated current will provide the capability to overcome any friction loads that might be encountered (See Fig. 6.6). This means that as much as 10% of rated current may be needed to achieve zero velocity under some conditions. Referring to Fig. 6.8 and setting $E_c = 0$, it can be seen that a position error will be required to produce this current, and this represents uncertainty in position output:

$$X_u = \frac{0.10 i_r}{K_{fx} K_{sa}} = \frac{K_q i_r}{10 A K_{vx}} = \frac{Q_r / A}{10 K_{vx}} \tag{4}$$

Thus, inaccuracies in static position due to anomalies in the forward loop are a function of maximum output velocity divided by the loop velocity gain.

Another type of position error, called following error, occurs when the command signal increases at a steady rate. For example, suppose that the position servo under consideration is asked to follow a programmed command signal, consisting of a series of ramps and holds. During the ramps, the servo must move at a steady velocity. Referring to Fig. 6.8, it can be seen that a finite error signal must be maintained to generate the needed velocity. The magnitude of this following error is:

$$X_f = \left[\frac{1}{K_{fx}}\right] E_e = \left[\frac{1}{K_{fx}}\right] \left[\frac{AV}{K_{sa} K_q}\right] = \frac{V}{K_{vx}} \tag{5}$$

Therefore, the following error is directly proportional to the velocity of the command signal, and is minimized by the use of high loop velocity gain.

The position errors discussed thus far have been those which can be minimized by a tight servoloop (high K_{vx}). To these errors must be added errors in the transducer mechanism. Even if infinite loop gain were achievable, the accuracy of the servo can be no more accurate than the transducer itself. There are many types of position transducers available, including potentiometers (both linear and rotary), and resolvers (rotary). The most important types of transducer inaccuracies are as follows:

Repeatability: When the load returns to a given position, will the transducer output always return to the same value, regardless of the direction of approach? Errors of this type can be caused by lost motion in the mechanical drive as well as by the transducer itself.

Resolution: The output of some transducers are not perfectly smooth. Instead they look like a staircase. Wirewound potentiometers are a classic example of this phenomenon.

Linearity: Sometimes it is necessary that the servo output be a very linear function of the command input. This might be important in a tracer application where both the command and feedback signals are generated by potentiometers, whose outputs must be matched to each other. Linearities on the order of $\pm 0.5\%$ of full scale are common, while $\pm 0.1\%$ or better is feasible. Sometimes the transducer mounting can create nonlinearities. Another source of transducer inaccuracy which is often overlooked is ripple. This is generally a characteristic of transducers excited by AC voltages, and is caused by imperfect filtering of the carrier signal. If the carrier frequency is selected properly, the response of the servo to the ripple can be minimized.

The discussion to this point has neglected the dynamic characteristics of the various components in the servoloop. These effects are very important. Usually, the most important dynamics are those of the load and drive system. Referring to Fig. 6.7, the open loop drive stiffness is K_t, the series combination of K_a and K_s. Note that K_a is comprised of the drive linkage itself and the stiffness of the oil trapped between the servovalve and actuator (see list of symbols). The mass of the load combines with the drive compliance to generate a second-order oscillatory mode, having a natural frequency $\omega_{lr} = \sqrt{K_t/M}$. Since the primary sources of damping for this "load resonance" are friction and hydraulic leakage, the damping ratio is usually small (typically 0.1).

Servovalve dynamics can also be important. These generally can be adequately described as second-order with a damping ratio of approximately 1.0. Either the load or the servovalve modes can be driven dynamically unstable by excessive loop gain. If the load reasonance has an appreciably lower natural frequency than the servovalve, a reasonably well-behaved closed loop response can be obtained with $K_{vx} = \zeta_{lr}\omega_{lr}$ (valid for damping ratios up to 0.5). Since ζ_{lr} is usually very difficult to estimate, the use of 0.1 is generally adequate for estimation purposes:

$$[K_{vx}]_{maximum} = 0.1\omega_{lr} \tag{6}$$

This will produce a closed loop step response which is basically first-order with a few low-amplitude oscillations superimposed on it. As was the case before dynamic components in the loop were considered, the time constant of this first-order response is approximately equal to $(1/K_{vx})$. If the servovalve possesses the lower natural frequency, a well-behaved response can be obtained with a loop gain of:

$$[K_{vx}]_{maximum} = 0.4\omega_{sv} \tag{7}$$

The closed loop response in this case will become more second-order in nature, with a natural frequency of approximately $0.5\omega_{sv}$ and a damping ratio of 0.5. The lower value of K_{vx} obtained from Eqs. (6) and (7) is the one which should be used in Eqs. (3), (4), and (5). The dominant equation determines the nature of the closed loop dynamic response.

In most cases, the dynamic response of the transducer is negligible compared to the servovalve and load resonance. Occasionally, however, the transducer dynamics dominate the loop. If this is the case and the dynamics can be described as second-order, the maximum value of K_{vx} should be approximately equal to the product of damping ratio and natural frequency (up to a maxium value of 1/2 the natural frequency). The dynamic response in this case will be rather complex and is beyond the scope of this chapter.

The tachometer feedback shown in Fig. 6.3 compares rotating speed with command setting. An error signal of appropriate magnitude and direction is sent to the pump controller to synchronize the system.

The velocity servo action shown in Fig. 6.3 is schematically identical to the position servo depicted in Fig. 6.7, except that the transducer measures velocity instead of position. Actually, velocity servos are more commonly used to control hydraulic motors than to control linear velocity, but this discussion will first consider the linear case, for consistency. Figure 6.9 shows an idealized block diagram for the velocity servo (dynamics effects again neglected).

The most obvious difference between this block diagram and Fig. 6.8 is that the velocity servo has no inherent integration in the forward loop. As

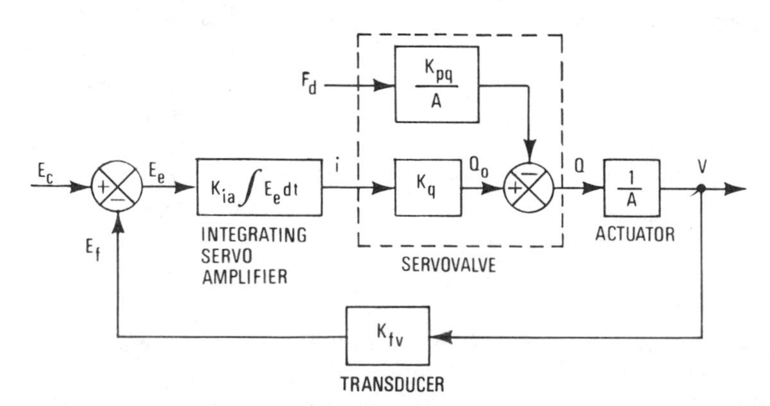

Figure 6.9 Idealized block diagram for a typical velocity servo. (Courtesy of Moog Inc., East Aurora, New York).

mentioned previously, such an integration is desirable to minimize static errors. Therefore, the integration in a velocity loop is generally provided electronically in the amplifier. This type of integration has a distinct advantage over the type inherent in the position servo because it is upstream of all forward loop anomalies, as well as upstream of where force disturbances enter the loop. Therefore, the static errors due to these effects are zero (the input current to the servovalve will continue to change until the error signal is driven exactly to zero).

The loop velocity gain is computed in the same manner used with the position servo:

$$K_{vv} = \frac{K_{fv}K_{ia}K_q}{A} \tag{8}$$

It still has the same units (sec^{-1}) but its physical significance is somewhat different. It now is a direct measure of how much acceleration the servo will generate in correcting a transient velocity error. If the command signal should be programmed so that there are some ramps with time (accelerations), the output velocity will follow the commanded velocity with a finite error. To generate an acceleration, $\dot{V}$, requires an error signal of $E_e = A\dot{V}/K_{ia}K_q$. The velocity error required to achieve this error signal is:

$$V_f = \frac{1}{K_{fv}} E_e = \frac{\dot{V}}{K_{vv}} \tag{9}$$

When loop dynamics are considered, the integrating velocity servoloop is identical to the position servoloop. Therefore, the maximum value of K_{vv} can be determined from Eqs. (6) and (7), and the form of the closed loop dynamic response is also identical.

In view of the foregoing discussion, the static errors in the velocity servo are determined almost entirely by the transducer characteristics. Typical velocity transducers are LVT's (linear), DC and AC tachometers, and magnetic pulse pickups. Ripple, linearity, and low-speed performance are generally the most important characteristics of these devices.

The third basic type of closed loop control system to be considered in this chapter is the force or pressure-control servo. Since force and pressure can be thought of interchangeably, the following discussion will use pressure exclusively. A schematic and block diagram for a typical pressure-control servo are shown in Figs. 6.10 and 6.11. Note that the block diagram of Fig. 6.11 is similar in form to the position servo depicted in Fig. 6.8, except that the disturbance input is now the velocity V. This velocity represents any movement of the object that the servo is pushing against, such as might occur during drilling or winching operations (the latter application normally uses a ro-

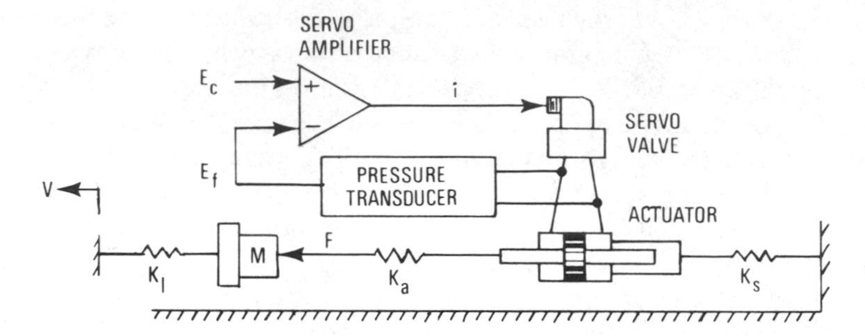

Figure 6.10 Idealized block diagram (schematic) pressure control servo. (Courtesy of Moog Inc., East Aurora, New York).

tary servo). Another difference from the position servo is that servovalve droop now represents an inner feedback loop. Neglecting this inner loop for the moment (its effects are small), the pressure loop has a loop velocity gain which is:

$$K_{vp} = \frac{K_{fp} K_{sa} K_q K_t}{A^2} \tag{10}$$

One source of static pressure errors is the droop due to hydraulic leakage in the servovalve (and actuator). The leakage flow due to output pressure

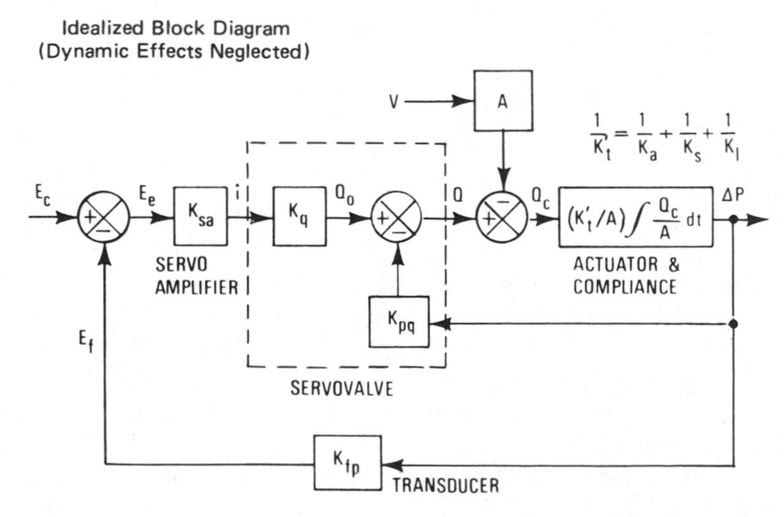

Figure 6.11 Typical pressure-control servo. Idealized block diagram (dynamic effects neglected). (Courtesy of Moog Inc., Easst Aurora, New York).

at a particular operating point is $K_{pq}\Delta P_{ss}$ (see Fig. 6.11). This leakage flow must be offset by a change in Q_o, which is generated by an error in output pressure equal to:

$$\Delta P_d = \frac{Q_o}{K_{fp}K_{sa}K_q} = \frac{Q_o K'_t}{A^2 K_{vp}} = \frac{K_{pq}K'_t}{A^2 K_{vp}} \Delta P_{ss} \tag{11}$$

From previous discussions, a conservative value of K_{pq} is $(0.05\ Q_r/P_s)$. Substituting this into Eq. (11), the result is:

$$\frac{\Delta P_d}{\Delta P_{ss}} = \frac{(Q_r/A)K'_t}{20(A\ P_s)K_{vp}} \tag{12}$$

Thus, there is always a pressure error present, which is a fixed percentage of the commanded pressure.

Another source of static errors in the pressure output will be servovalve and drive system anomalies which can amount to 10% of rated current, as previously discussed. This is the amount of input current that might be required to achieve zero flow (and therefore a steady value of ΔP) under some conditions. Referring to Fig. 6.11 and setting $E_c - 0$, it can be seen that a pressure error will be required to produce this current, and this represents uncertainty in output pressure:

$$\Delta P_u = \frac{0.10\ i_r}{K_{fp}K_{sa}} = \frac{K'_t K_q i_r}{10A^2 K_{vp}} = \frac{(K'_t/A)(Q_r/A)}{10\ K_{vp}} \tag{13}$$

The numerator of this expression is the maximum rate of change of pressure that the servo is capable of generating. Another type of error occurs for those applications which require pressure to be held in the presence of load velocities. This error can be calculated by observing that Q_c must be zero if a steady-state output pressure is to be achieved. Referring to Fig. 6.11 and setting Q_c and E_c equal to zero, there results:

$$\Delta P_f = \frac{VA}{K_{fp}K_{sa}K_q} = \frac{K'_t}{AK_{vp}} V \tag{14}$$

The primary dynamic element in the loop is the servovalve. Of course, there is a load resonance with $\omega_{lr} = \sqrt{K_t + K_l)/M}$, but this mode is accompanied by a complex pair of zeros which have a natural frequency somewhat below ω_{lr}. This means that high loop gains cannot drive the load resonance unstable. In fact, high loop gain tends to improve the damping ratio of the load resonance and submerge its effects in the closed loop response. Therefore, the maximum usable value of K_{vp} can be determined from Eq. (7), and the closed loop dynamic response will be essentially second-order with a natural frequency of $0.5\omega_{sv}$ and a damping ratio of 0.5.

Additional static errors occur due to the transducer itself. The most common types of transducers used are diaphram pressure transducers or load cells (force transducers). The actual transducing element in both devices is usually a strain gauge. Important performance characteristics are repeatability and linearity. In addition, the low output levels of strain gauges require high-gain signal conditioning, which can lead to amplifier drift problems and high noise susceptibility.

Performance Summary

For convenience, the foregoing performance equations have been summarized in Table 6.1. To use the Table, the following steps are necessary.

1. Select a transducer, being careful that its performance is well within the limits allowable for the servo as a whole. Consider how it will be mounted and what quantity it will actually measure (remember that the load stiffness of a position servo is highly dependent upon transducer location).
2. Estimate drive stiffness so that the natural frequency of the load resonance can be calculated. Also determine the equivalent natural frequency of the servovalve (can use frequency for 90° of phase lag from a frequency response). Make sure that the natural frequencies of the transducer are well above both of these values if possible.
3. Calculate maximum usable loop gains and determine the closed loop dynamic response.
4. Compute the servoloop inaccuracies appropriate to the application at hand. Bear in mind that transducer inaccuracies add directly to the values determined from the table.

For applications involving rotary motion instead of linear motion, Figs. 6.7–6.11 and all the performance equations can very easily be converted by simply substituting motor displacement for piston area, angular motion for linear motion, torque for force, moment of inertia for mass, and angular spring rates for linear rates. Table 6.2 summarizes the results of these substitutions. If a particular application involves gearing between the motor and load, Table 6.2 can still be used by reflecting all parameters back to the motor output shaft.

Conclusions

The performance formula given in Tables 6.1 and 6.2 make it relatively easy for engineers to estimate the performance they are likely to achieve if they utilize an electrohydraulic servo in a particular control system application. However, several words of caution are in order.

1. The ground rules and qualifying assumptions stated in the text are very important.

Table 6.1 Performance Formulas – Linear Servos

Performance parameters	Position servo	Velocity servo	Force servo
Maximum usable loop velocity gain and closed loop dynamic response	$K_{vx}, K_{vv} = 0.1\omega_{lr}$	(Closed loop response first-order; time constant $= 1/K_{vx}$)	$K_{vp} = \omega_{sv}$
	$K_{vx}, K_{vv} = 0.4\omega_{sv}$	(Closed loop response second-order; natural frequency $= 0.5\omega_{sv}$, damping ratio $= 0.5$)	Closed loop natural frequency $= 0.5\ \omega_{sv}$ Closed loop damping ratio $= 0.5$
	Note: Select path which gives lowest value of loop gain		
Maximum output uncertainty[a]	$X_u = \dfrac{Q_r/A}{10\ K_{vx}}$	Essentially zero	$\Delta P_u = \dfrac{(K'_t/A)(Q_r/A)}{10\ K_{vp}}$
Output stiffness[b]	$\dfrac{F_d}{X} = 20\ K_{vx} \dfrac{A\ P_s}{Q_r/A}$	Essentially infinite	Not applicable
Leakage errors	Not applicable	Not applicable	$\dfrac{\Delta P_d}{\Delta P_{ss}} = \dfrac{(Q_r/A)\ K'_t}{20(A\ P_s)K_{vp}}$
Following errors	$X_f = \dfrac{V}{K_{vx}}$	$V_f = \dfrac{V}{K_{vv}}$	$\Delta P_f = \dfrac{K'_t}{A\ K_{vp}} V$

[a]Transducer inaccuracies directly add to these values.

[b]Any drive system compliance outside the loop will reduce this term (see text).

(Courtesy of Moog, Inc., Buffalo, New York)

Table 6.2 Performance Formulas—Rotary Servos

Performance parameters	Position servo	Velocity servo	Force servo
Maximum usable loop velocity gain and closed loop dynamic response	$K_{v\theta}, K_{v\Omega} =$ $0.1\omega_{lr}$ $0.4\omega_{sv}$	(Closed loop response first-order; time constant $= 1/K_{v\theta}$) (Closed loop response second-order; natural frequency $= 0.5\omega_{sv}$, damping ratio $= 0.5$)	$K_{vp} = 0.4\omega_{sv}$ Closed loop natural frequency $= 0.5\ \omega_{sv}$ Closed loop damping ratio $= 0.5$
	Note: Select path which gives lowest value of gain		
Maximum output uncertainty[a]	$\theta_u = \dfrac{2\pi Q_r/D}{10 K_{v\theta}}$	Essentially zero	$\Delta P_u = \dfrac{(2\pi G'_t/D)(2\pi Q_r/D)}{10\ K_{vp}}$
Output stiffness[b]	$\dfrac{T_d}{\theta} = 20\ K_{v\theta}\ \dfrac{DP_s/2\pi}{2\pi Q_r/D}$	Essentially infinite	Not applicable
Leakage errors	Not applicable	Not applicable	$\dfrac{\Delta P_d}{\Delta P_{ss}} = \dfrac{(2\pi Q_r/D)\ G'_t}{20(D\ P_s/2\pi)K_{vp}}$
Following errors	$\theta_f = \dfrac{\Omega}{K_{v\theta}}$	$\Omega_f = \dfrac{\dot{\Omega}}{K_{v\Omega}}$	$\Delta P_f = \dfrac{G'_t}{(D/2\pi)K_{vp}}\Omega$

[a]Transducer inaccuracies directly add to these values.

[b]Any drive system compliance outside the loop will reduce this term (see text).

(Courtesy of Moog, Inc., Buffalo, New York)

2. The formulae are designed to be conservative for most applications, but there are no guarantees that they are conservative in every case.
3. Do not overlook the possibility that simple open loop control may be adequate for a given application. In applications requiring a full-time human operator, such as on mobile equipment, closed loop control may offer no advantage whatsoever. Human beings are extremely adaptable and can often do a better job of closing the loop than a servo system can.

If electrohydraulic controls appear to be well suited to an application, but the formulae given in this chapter indicate that performance will not be adequate, do not give up! There are many techniques which can often be utilized to improve performance well beyond what is achievable with the simple servoloops considered here. Get help from an experienced servo engineer. In the world of closed loop control systems, almost anything is possible. The only real limitations are time and money.

6.3. CONTROL VALVES

6.3.1. General Characteristics

Manual Controls

Pump Controls The manual *linkage system* shown in Fig. 6.4(a) has been incorporated into many industrial applications in addition to the basic use as a control for a piston pump. The electrohydraulic servovalve has replaced many of these manual applications because of the flexible control capabilities of the electrical system.

Torque Generator Power Steering Units The torque generator power steering unit shown in Fig. 6.12(a) is mechanically linked to the actuated device. It can be mounted directly on steering or turning shaft.

This torque generator incorporates a hydraulic motor and a rotary servovalve in a common housing [Fig. 6.12(b)]. The torque *boost* is produced by hydraulic pressure from an external source operating on the hydraulic motor as controlled by the servovalve which provides a manual input signal. The input shaft and output shaft are *mechanically* linked so that manual torque can be delivered when the pump is temporarily inoperative [Fig. 6.12(c)].

The torque generator has applications other than power steering, such as large gate valve operators and manual winch operation.

Fully Fluid Linked Power Steering Units Fully fluid linked power steering control units are somewhat different than the torque generator of Fig. 6.12(a). A fully fluid linked power steering control system does not require a mechanical connection between the steering unit, the pump, and the steering cylin-

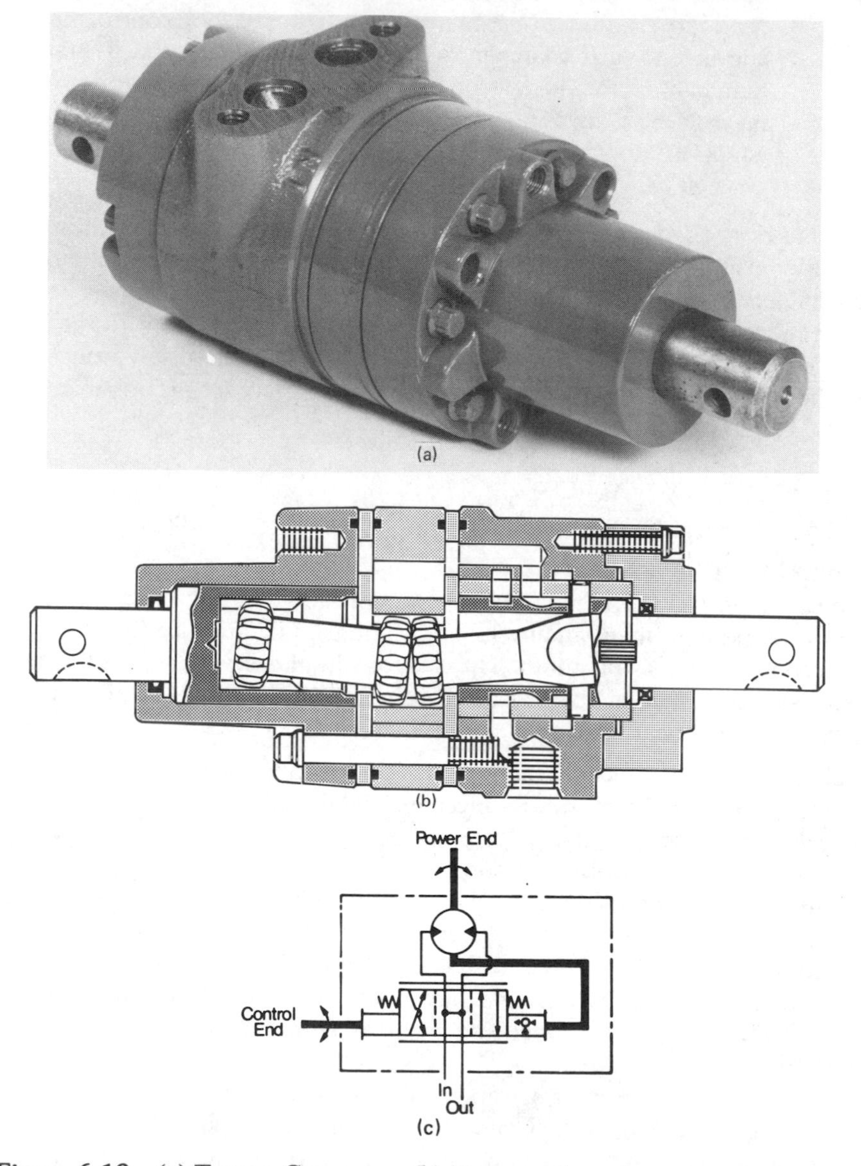

Figure 6.12 (a) Torque Generator. (b) Cross section view torque generator. (c) Symbol for torque generators. (Courtesy of Eaton Corp., Hydraulics Div., Eden Prairie, Minnesota).

ders. The unit consists of a manually operated directional control valve and servo feedback meter element in a single body. It is used principally for fluid linked power steering systems but it can be used for some servo-type applications or any application where visual positioning is required. The close-coupled, rotary action valve performs all necessary fluid directing functions with a small number of moving parts. The manually actuated valve is coupled with the mechanical drive to the meter gear. The control is lubricated and protected by the power fluid in the system and can operate in many environments.

In some installations there is no need for a load reaction. A nonload reaction steering unit blocks the cylinder ports in neutral, holding the axle position whenever the operator releases the steering wheel (Fig. 6.13).

When a load reaction is desirable, the circuit can be changed per Fig. 6.14. A load reaction steering unit couples the cylinder ports internally with the meter gear set in neutral so that the axle forces can return the steering wheel to its approximate original position when the operator releases the wheel after completing a turn (like automotive steering). The cylinder system used with load reaction units must have equal oil volume displaced in both directions. The cylinders should be a parallel pair (as shown) or one double rod end unit. Do not use with a single unequal area cylinder system.

A complete load sensing system diagram is shown in Fig. 6.15 using a fixed displacement (open center) power supply. The circuit can use a pressure compensated (closed center) power supply as shown in Fig. 6.16. Figure 6.17 shows the load sensing steering system with flow and pressure compensated (load sensing) power supply.

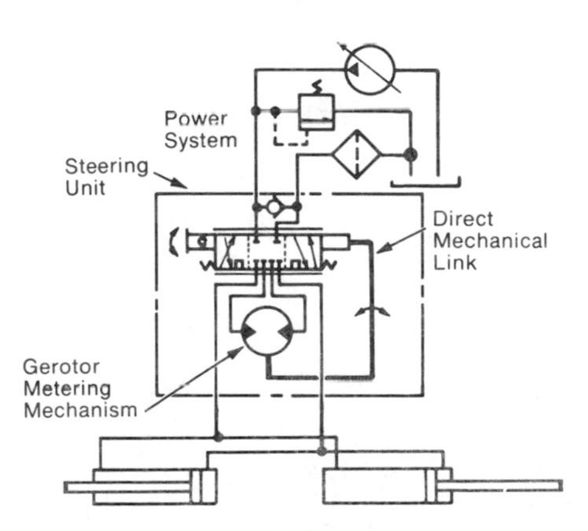

Figure 6.13 Closed center system nonload reaction circuit. (Courtesy of Eaton Corp., Hydraulics Div., Eden Prairie, Minnesota).

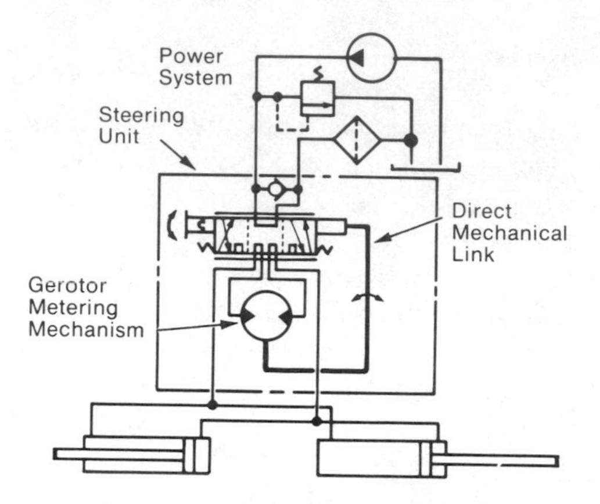

Figure 6.14 Open center system load reaction circuit. (Courtesy of Eaton Corp., Hydraulics Div., Eden Prairie, Minnesota).

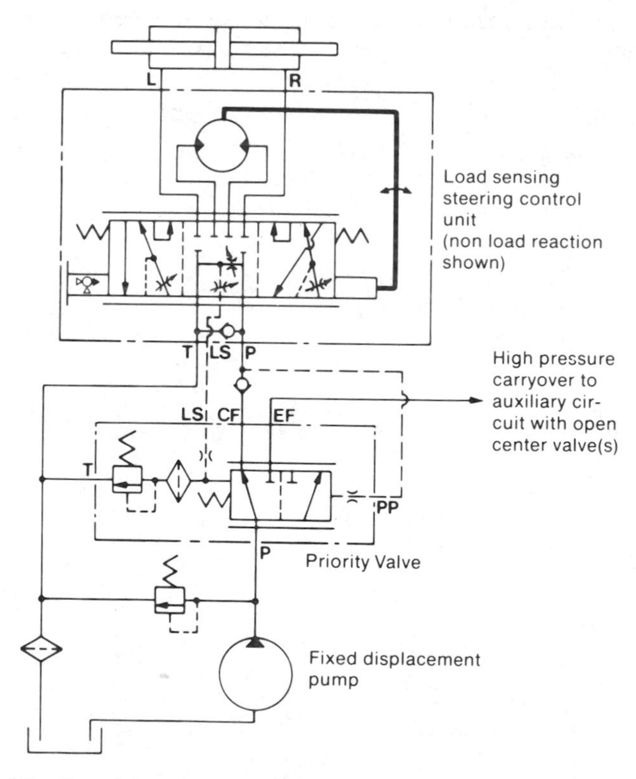

Figure 6.15 Load sensing steering system with fixed displacement (open center) power supply. (Courtesy of Eaton Corp., Hydraulics Div., Eden Prairie, Minnesota).

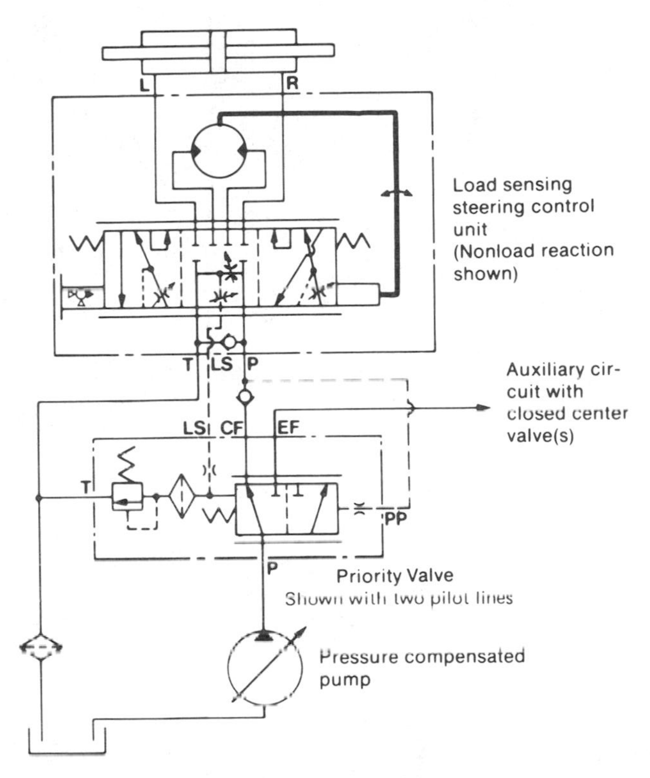

Figure 6.16 Load sensing steering system with pressure compensated (closed center) power supply. (Courtesy of Eaton Corp., Hydraulics Div., Eden Prairie, Minnesota).

The use of a load sensing steering unit and a priority valve in a normal power steering circuit offers the following advantages:

1. It provides smooth pressure compensated steering because pressure variations in the steering circuit do not affect steering response or maximum steering rate.
2. It provides true power beyond system capability by splitting the system into two independent circuits. Pressure transients are isolated in each circuit. Only the flow required by the steering maneuver goes to the steering circuit. Flow not required for steering is available for use in the auxiliary circuits.
3. It provides reliable operation because the steering circuit always has flow priority.

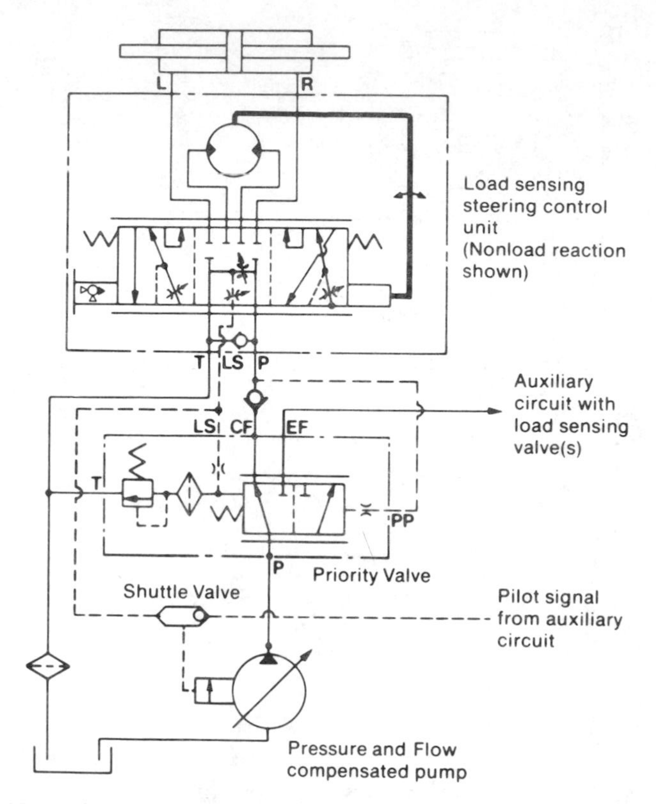

Figure 6.17 Load sensing steering system with flow and pressure compensated (load sensing) power supply. (Courtesy of Eaton Corp., Hydraulics Div., Eden Prairie, Minnesota).

Electric Controls

Torque Motor A common bidirectional, proportional electromechanical (EM) interface device is the torque motor (Fig. 6.18). The term torque *motor* may be misleading in that the armature of the motor moves (rotates) only a few minutes of an arc, but the torque created on the armature is proportional in amplitude and direction to the value of current in the coils. Torque motors are generally compact, have high torque-per-unit current, and have dynamic response that is more than adequate for most hydraulic applications (usually to several hundred Hertz). The torque motor response is the key to satisfactory valve actuation time and the overall response of the system.

Pilot Stages Most electrohydraulic remote controls include a pilot stage between the electromechanical interface device and the hydraulic component being controlled. The pilot stage is sometimes called an hydraulic amplifier as it is supplied with hydraulic power (pressure P and return R), and this power is modulated, or controlled, by the mechanical output of the EM device. The result is a hydraulic pressure which can develop the high force needed to control the hydraulic equipment. The pressures involved may be less than 3000 psi because of limitations of available pumps, motors, valves, and hydraulic plumbing. With the advent of higher pressure pumps, valves, and particularly better hydraulic hose, the pressures may reach 4000 psi and above. The higher pressures result in smaller and lighter weight components and systems which can save energy in many industries and particularly on airborne equipment.

A typical pilot stage is a double nozzle and flapper (Fig. 6.19). Two fixed orifices and two variable orifices are arranged in a Wheatstone bridge circuit. Flapper motion causes the output port pressures A or B to swing one way or the other. A second hydraulic stage is used to increase flow capabilities with some reduction in response characteristics.

Other types of pilots include jet-type [Fig. 4.15(c)] where the momentum of a jet of fluid that falls on a receiver is controlled, and small sliding spools. Generally jet-type pilot stages do not work well at low temperatures

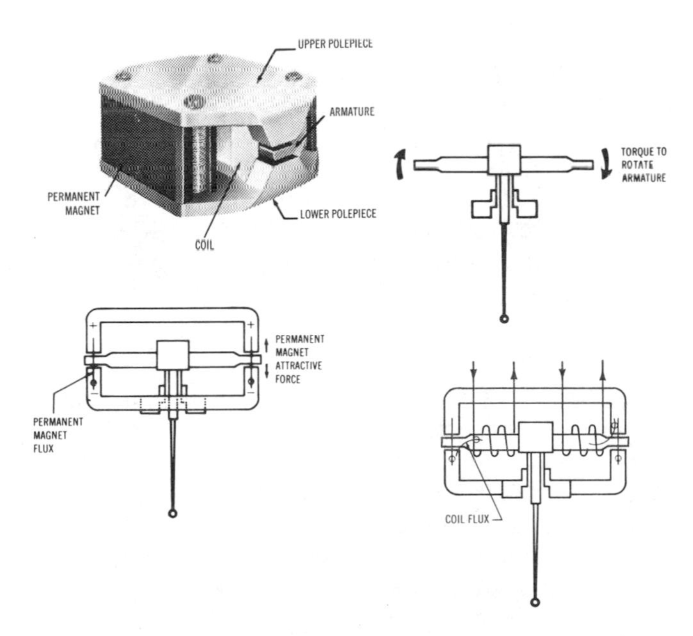

Figure 6.18 Torque motor. (Courtesy of Moog Inc., East Aurora, New York).

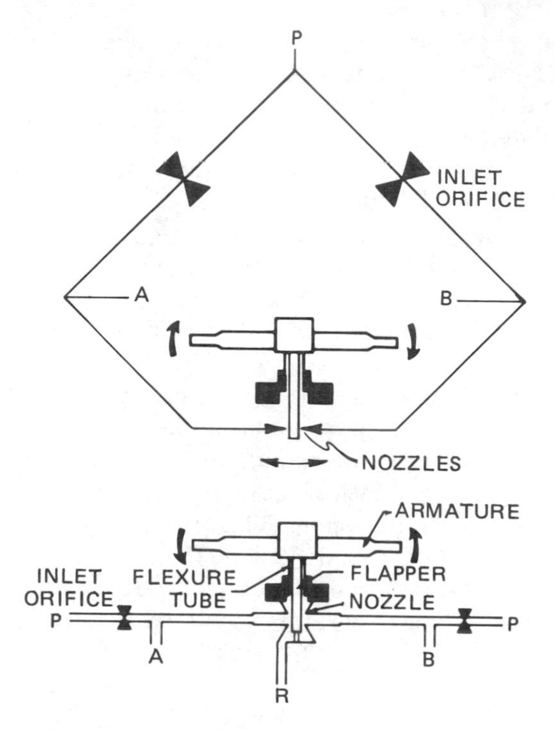

Figure 6.19 Double nozzle and flapper pilot stage. (Courtesy of Moog Inc., East Aurora, New York).

(less than 0°F). Sliding spool valves are not preferred for direct control by a torque motor or other EM devices as spool friction (which can become severe with fluid contamination) can represent a sizable portionof the maximum driving force.

6.3.2. Open and Closed Center Valves

Spool configurations for proportional valves operating on constant pressure systems are either open center or closed center.

The term closed center is common in all areas of hydraulic terminology. The degree of seal for a closed center spool is usually very small as related earlier in this chapter because it creates a dead band. The dead band slows up response of the valve when a signal is provided. Open center is often considered all ports connected to each other and to tank. A typical servo system usage is shown in Fig. 6.20 which differentiates to the extent that pressure is blocked even though the cylinder ports are connected to tank.

In this circuit the open center spool has both control ports open to return when the spool is centered. The pressure lands are overlapped so the up to 10% to 20% of spool travel is necessary to open a cylinder port to pressure. This spool configuration is always used to operate a man lift as the open center insures operation of load holding valves at the lift cylinders (which are required by OSHA/ANSI).

Open center spools may have shaped control slots so that flow from the valve builds up more gradually when the valve is near its center position.

Figure 6.21 shows a typical curve of flow versus spool position as movement is initiated by application of electrical energy to the direct current solenoid.

Closed center spools give very linear flow versus spool position. By design, closed center spools can be nonlinear if desired.

This linearity does not give very normal "feel" for a human operator, but it does produce good closed loop servocontrol.

6.3.3. Proportional Operation

Proportional valves have either of two different mechanizations: (1) spring centered spools, or (2) "free floating" spools with mechanical feedback. Spring centered spools move to a position corresponding to the pressure produced by the pilot stage balancing the compressed spring load, presuming there is negligible spool friction. This valve configuration requires good linearity and

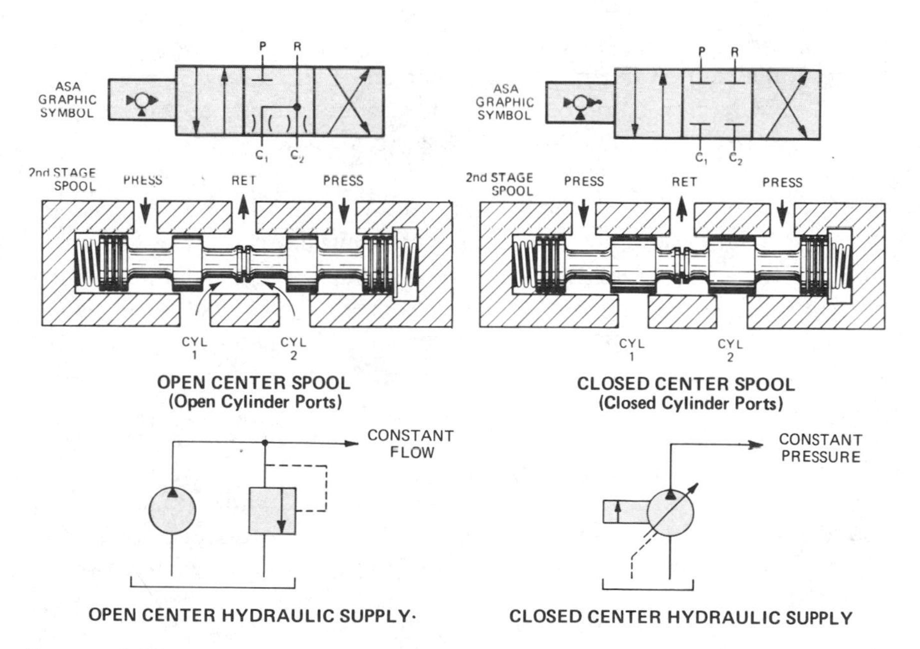

Figure 6.20 Servo valve configurations. (Courtesy of Moog Inc., East Aurora, New York).

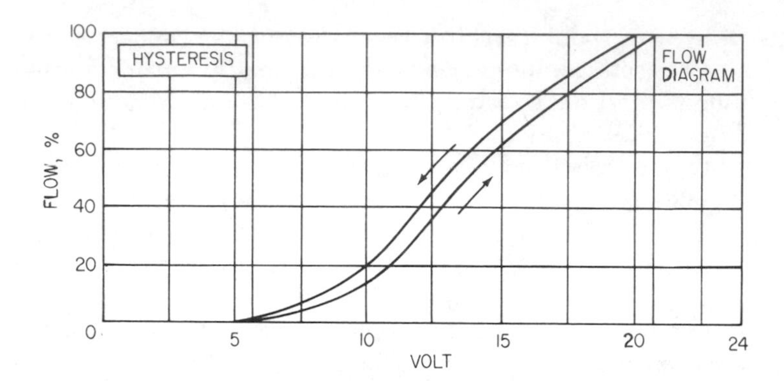

Figure 6.21 Typical flow curve for spool valve. (Courtesy of HIP/Nichols, Sturtevant, Wisconsin).

repeatability from the pilot, which can be a problem over wide ranges of temperature and pressure.

The preferred technique is to allow the spool to move freely (small detent centering springs are usually included to insure centered spool with pressure off). Only small pilot pressures are normally necessary to move the spool. Movement of the spool creates a feedback torque through the feedback wire that counteracts the electrical input torque, so the spool stops at a position proportional to the value of the input current (Fig. 6.22).

This valve configuration has the advantage of being able to build up large pressures on the spool, if necessary, to move it.

6.3.4. Open Loop Proportional Control

Control of a proportional valve such as the one just described is a simple task. The maximum electrical power required is about 1/8 W, which can be supplied through a potentiometer from a 12 V battery. These valves can also be fully controlled by a similar arrangement from a 4 V miner's cap battery. The valve can be made intrinsically safe by adding Zener diodes across the torque motor coils.

The circuit is a Wheatstone bridge, with the potentiometer forming two adjacent arms of the bridge (Fig. 6.23).

Valves can be stacked, and the stack connected with a single pressure and return line. When the load demand regulator is used, one additional small line per stack is necessary to connect to the pump pressure regulator (similar to that shown in Fig. 7.3).

The single lever control stations can also be stacked in a family in-line arrangement. Two-axis joy stick controllers are also available, as are multi-axis (up to six), hand-held controllers.

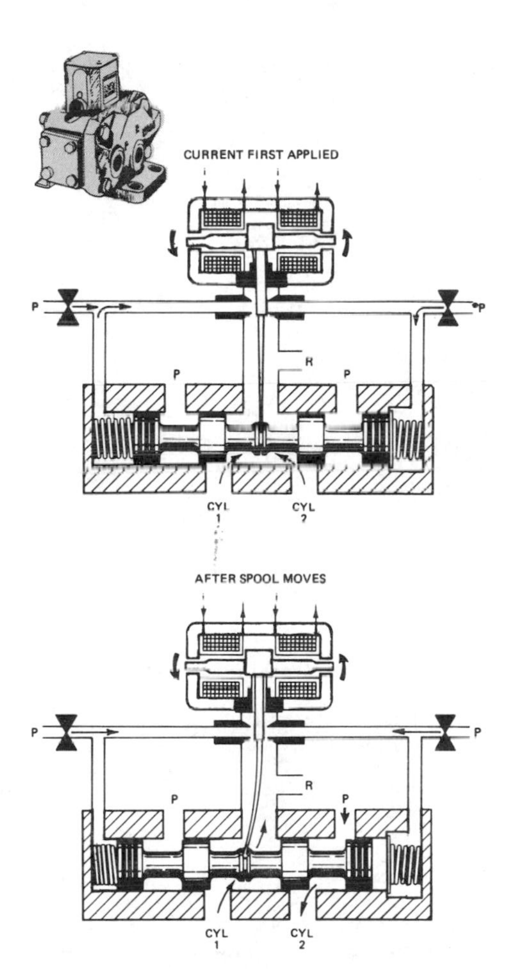

Figure 6.22 Operation of proportional valve. (Courtesy of Moog Inc., East Aurora, New York).

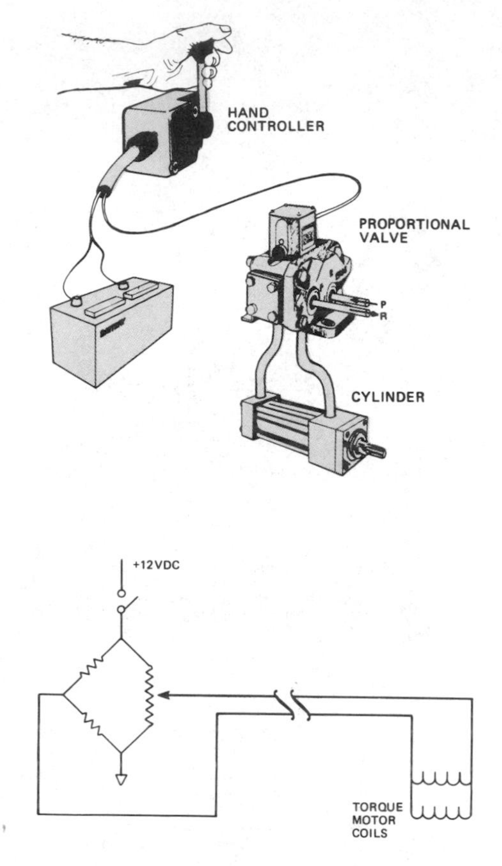

Figure 6.23 Open loop proportional control. (Courtesy of Moog Inc., East Aurora, New York).

6.4. PUMP CONTROLS

Proportional pumps and motors provide a useful actuator choice. Hydrostatic controls using variable displacement pumps, or motors, or both, are a type of hydraulic component for which remote control is of significant importance (Fig. 6.24).

Conventional hydrostatic pump and motor controls use one or two pistons to move the variable stroking mechanism, a valve to control the pistons, and a mechanical linkage to position the valve (Fig. 6.4).

Electrical remote controls for hydrostatic pumps and motors are provided by a proportional valve mechanism like that previously described [Fig.

6.4(b)]. In this case, however, it is necessary to have the position of the pump stroking mechanism proportional to the electrical input, so a lever and spring connect from the stroking mechanism to the torque motor. When an electrical signal is applied, the valve will act. This causes the stroking mechanism to move until it reaches a displacement where the feedback spring just counteracts the electrical input torque. At this point the valve has moved back to near its centered position.

6.5. TYPICAL EXAMPLES OF SERVO SYSTEMS

6.5.1. Open Loop Steering Control

The advantages of electrical remote control of hydraulic equipment often result from the ease with which electrical signals can be manipulated.

The steering/propulsion control for a track-type crawler vehicle (Fig. 6.25) offers a typical example. The tracks are individually driven by hydrostatic pumps and motors. A single joy stick hand controller gives simultaneous forward/reverse and right/left control.

The electrical connections to accomplish this make use of dual coil torque motors. A coil of each pump controller is driven from the potentiometer that senses forward/reverse joy stick motion. The other coils are driven differentially from the potentiometer that senses side-to-side motion.

The crawler will turn in minimum space by moving the joy stick straight to the right or left. The track velocity associated with turning can be set at

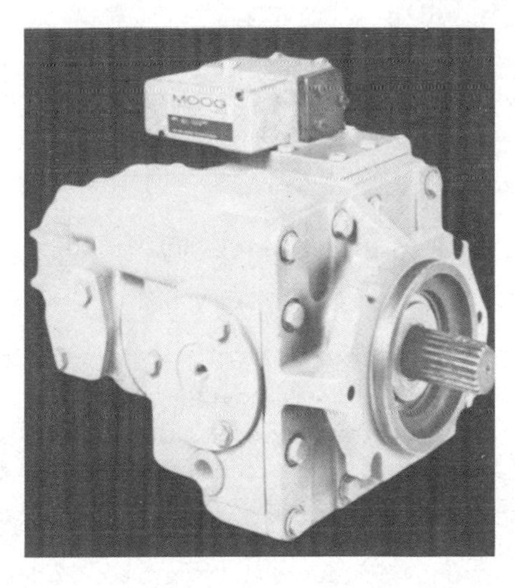

Figure 6.24 Piston pump with electric controller. (Courtesy of Moog Inc., East Aurora, New York).

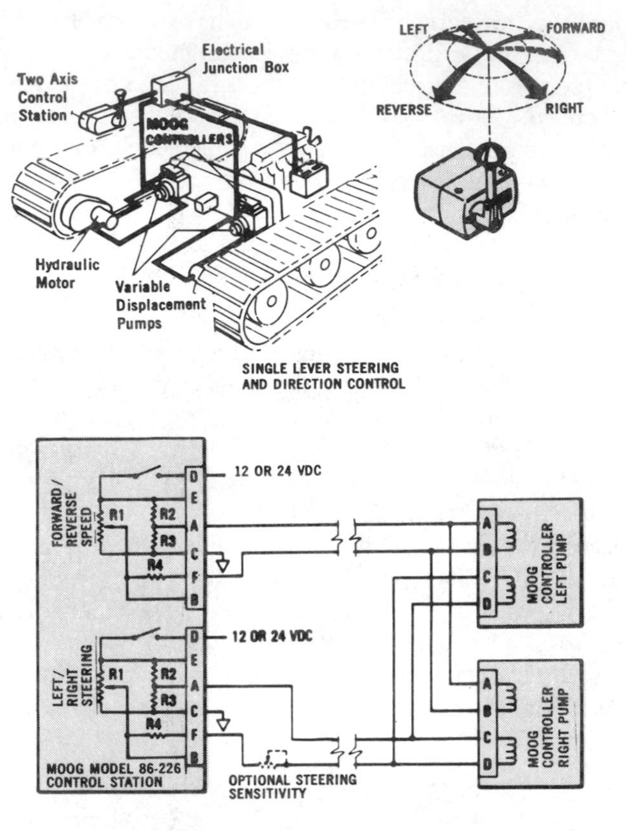

Figure 6.25 Crawler vehicle steering control. (Courtesy of Moog Inc., East Aurora, New York).

a lower value than the forward/reverse velocity by adding a dropping resistor in the steering circuit.

6.5.2. Pump/Motor Crossover Control

When a variable displacement motor, together with a variable displacement pump, are used to give an extra wide speed range, the signals to the pump and motor controls can be supplied by a simple crossover circuit (Fig. 6.26).

With zero speed setting, the pump is at zero stroke and the motor at maximum displacement. As the speed setting is advanced, the pump displacement picks up by supplying current to the pump controller. When the pump is fully stroked, further increase in speed is achieved by destroking the motor. The motor contains physical stops that prevent displacements less than about

40% maximum. Adjustment of the relative sensitivities of the pump and motor control currents minimizes the inherent nonlinearity of the inverse motor stroking control.

The circuitry to accomplish this is contained on a small printed circuit card that fits inside a standard hand control station.

6.5.3. Antistall Control

Another practical adaptation of electrical controls is the provision for antistall. Hydraulic controls are relatively fast and *stiff* such that it is frequently possible to overload the prime mover. This annoyance – or danger – can be avoided by incorporating override of the load demanding control by a signal that senses overload of the prime mover. If the prime mover is an electric

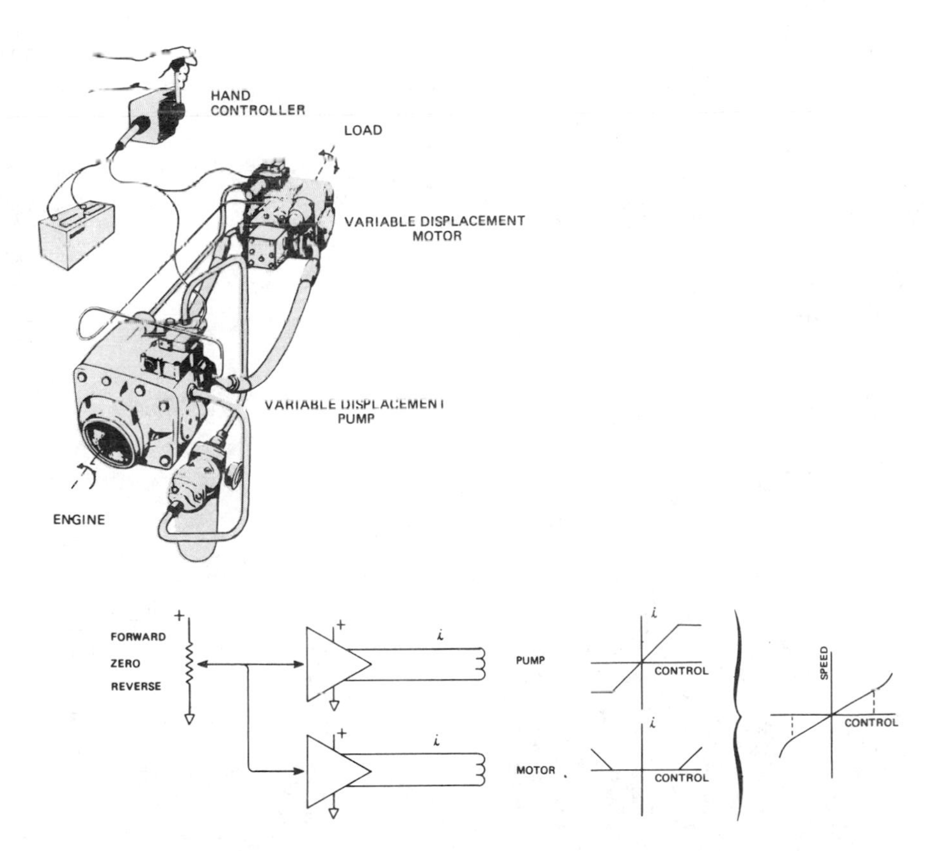

Figure 6.26 Pump/motor crossover control. (Courtesy of Moog Inc., East Aurora, New York).

motor, then overload can be sensed by monitoring the motor current. When the prime mover is an internal combustion engine, then the overload sensing scheme must provide for the normal engine speed setting. The arrangement shown in Fig. 6.27 does this by comparing actual engine speed (measured by tachometer) with the engine speed setting (potentiometer on the throttle). If the engine speed is above the throttle setting, then full velocity speed command is available. This is provided by full supply voltage to the vehicle speed control pedal.

The output from this pedal setting supplies current to the pump stroker. A low pass filter is located in series with the speed command so that rapid speed changes (from an overzealous operator) are avoided.

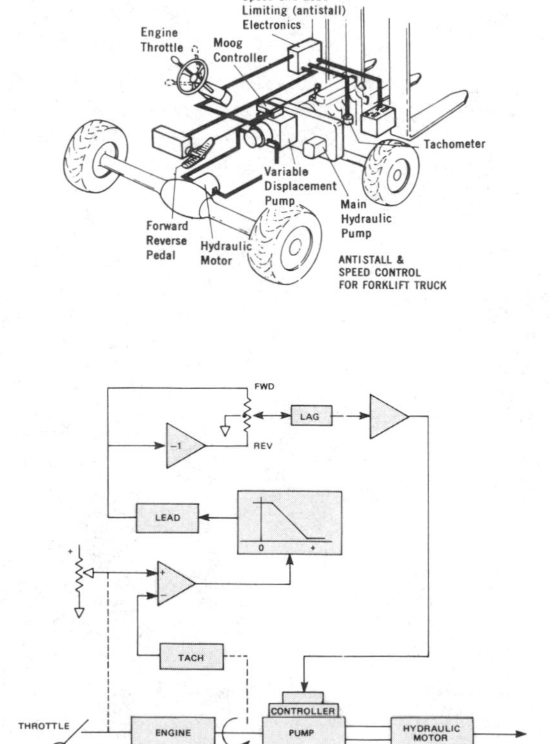

Figure 6.27 Antistall control. (Courtesy of Moog Inc., East Aurora, New York).

If the engine slows down due to overload (vehicle speed command is excessive for the terrain and work being performed), the voltage to the speed control is reduced. This override circuit has a lead network to offset the lag located between the speed setting and the pump controller. This lead compensation allows the pump controller to respond quickly should an overload exist.

Other functions often required by specific machinery served by hydrostatic drives that can be readily provided by electric remote control include horsepower, engine overspeed, and drive overtorque limiting. In each case, electrical transducers are added to sense the appropriate variable, and the transducer signal is used to control or override the normal hydrostat stroke signal.

6.5.4. Multichannel Control

Each of the control circuits discussed have been single functions. Most vehicles and equipment require combinations of functions such as steering + crossover + antistall + overspeed limiting + overtorque limiting. Some vehicles and equipment have several entirely different controls including, perhaps, a mixture of valve and hydrostatic controls. If remote control is used, it makes sense to provide control for all functions. The operator (Fig. 6.28) must be able to control at least major functional machine movements from his vantage point. Also, it may be desirable to provide portable control stations such as that shown in Fig. 6.30 which can be connected with a suitable umbilical cord. In this way the operator can move to the most advantageous location, on or off the equipment. The ultimate in portability and operator freedom is radio control (Fig. 6.29). A nonconductive data link (NDL) used also in vehicles that operate near high tension electrical wires consists of a multichannel radio control that is designed for safe and reliable operation of commercial vehicles and equipment.

Command signals are derived, as before, from potentiometers connected to levers that are manipulated by the operator. The analog potentiometer signals are converted to binary digital signals. These signals, together with a number of on/off switch signals for setting other machine functions, are sampled at a fast rate (20 times each second). This sampling, or multiplexing, produces a chain of pulses that carry commands for the proportional functions and commands for the on/off functions. An operator identification code is superimposed on the multiplexed signal for safety, and also to allow several pieces of radio controlled equipment to operate together without false or ambiguous commands. A battery powered RF transmitter broadcasts the command signal to a receiver located on the equipment being controlled. The receiver contains logic circuitry that unscrambles the multiplexed command information. This is supplied to the amplifiers and relays that drive the various controls in the vehicle or equipment.

Figure 6.28 Boom truck with personnel basket. (Courtesy of Moog Inc., East Aurora, New York).

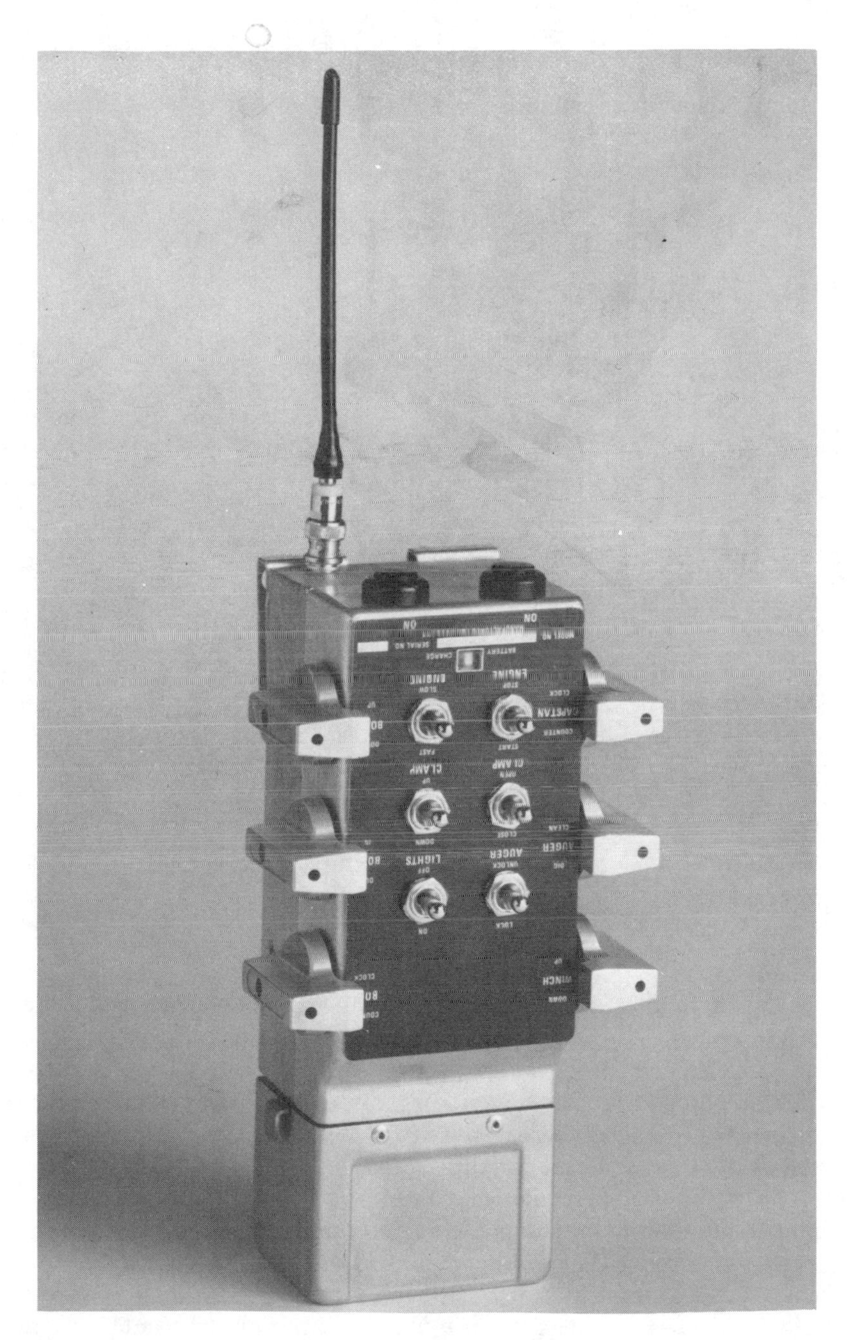

Figure 6.29 Radio controller. (Courtesy of Moog Inc., East Aurora, New York).

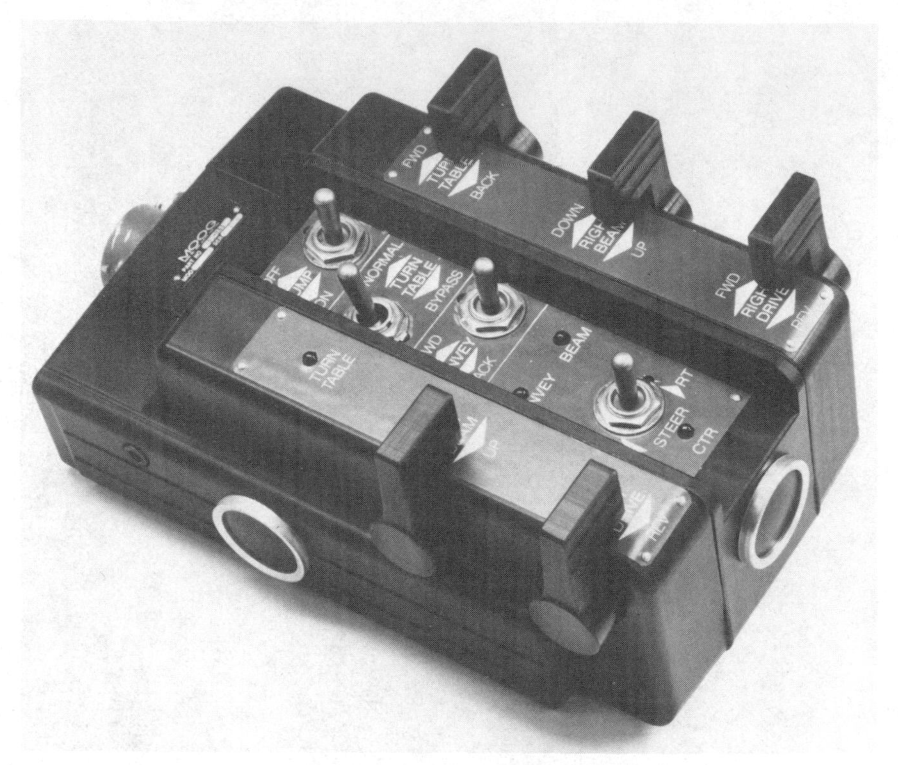

Figure 6.30 Belt mounted controller. (Courtesy of Moog Inc., East Aurora, New York).

An adaptation from the nonconductive data link controller that is advantageous when many channels are involved, but when radio control is not needed or desired, is the signal multiplexing. Direct electrical connections to provide numerous channels of control require many wires. If the operator controller is hand held (tethered), the size of the connecting cable can become unwieldly. A good solution is multiplexing.

The multiplexing circuitry is located on a printed circuit card that can be contained inside a multichannel hand controller. The cable that connects the hand controller to the equipment consists of three small wires. Even with flexible, steel armored sheathing, the cable is only 5/16 in. diam. This cable is easily handled by the operator, or it may be wound or unwound by a spring driven take-up reel. The controls can be grouped in a console as shown in Fig. 6.31.

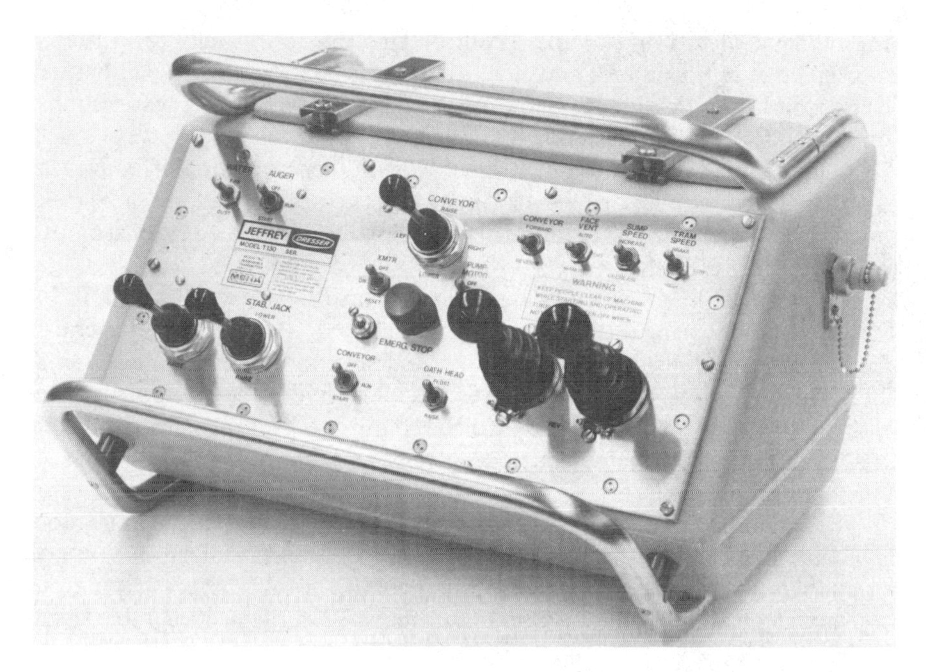

Figure 6.31 Custom control station. (Courtesy of Moog Inc., East Aurora, New York).

6.6. SUMMARY

The practical era of servo control systems and electrical remote controls for hydraulic equipment is here. Remote controlled equipment is in daily use, landmark installations have proven the dependability and reliability, and new systems are being designed, evaluated, and placed into production at a growing rate.

The machine tool industry has adapted the controls discussed in this chapter to the power transmission systems employed with industrial robots, automated machining systems, painting, welding, flame cutting, and other similar jobs using both servovalves and systems and servocontrolled variable displacement pumps and motors (both linear and rotary). Manufacturing processes include tape controlled machines which can control position, velocity, force, and all usual mechanical motions associated with the manufacturing process.

Use of servo systems and electrical remote control is not a gimmick. Nor are they necessarily chosen because of improved performance (better smoothness, wider range of speed control, etc., due to this feature). Instead, the choice of servo systems and electrical remote control is usually either: (1) an economic justification, as in the case of most manufacturing installations, or (2) a safety related design solution.

Economic justification can result from higher operating effectiveness (increased productivity), or reduced manpower to accomplish the same task, or reduced damage to the equipment or payload due to the available smooth operation and effective speed control.

Safety can be provided by removing the operator from a hazardous location, by eliminating all electrically conductive connections to the operator (radio control), or by preventing dangerous load conditions from occurring.

In some cases electrical remote control of hydraulic equipment is required as it is the only manner by which the job can be done. Electrical control may be the best way to interface program data from source to the resulting machine cycle.

Computer-aided design and computer-aided manufacturing has this tool available to provide any movement pattern, speed control, acceleration and deceleration control, torque control, and combinations thereof in an accurate and dependable program which can emanate from any acceptable signal source.

Whatever the reason, these hydraulic control techniques should be considered during the design of every piece of hydraulically powered or controlled equipment.

LIST OF SYMBOLS

General

E_c	Command signal	*V*
E_e	Error Signal	*V*
E_f	Feedback signal	*V*
i	Servovalve drive current	*ma*
i_r	Rated current; i to obtain Q_r	*ma*
K_{fp}	Pressure transducer gain	*V/psi*
K_{ia}	Gain of integrating servoamplifier	$\frac{ma/sec}{V}$
K_p	Servovalve null pressure gain (blocked load)	*psi/ma*
K_{pq}	Servovalve droop; K_{pq} K_q/K_p	*cis/psi*
K_q	Servovalve flow gain (no load)	*cis/ma*
K_{sa}	Gain of proportional servoamplifier	*ma/V*

K_{vp}	Velocity gain of pressure loop *sec*$^{-1}$
ΔP	Differential pressure across piston or motor *psi*
ΔP_d	Error in pressure due to droop (K_{pq}) *psi*
ΔP_f	Pressure following error *psi*
ΔP_{ss}	Steady-state ΔP at any particular operating point *psi*
ΔP_u	Uncertainty in pressure due to anomalies in the servovalve and drive system *psi*
P_s	Net supply pressure *psi*
Q	Flow output of servovalve; neglecting compliance, $Q = VA$ or $(D/2\pi)\Omega$ *cis*
Q_o	Q under no-load conditions *cis*
Q_r	Rated value of Q_o at a particular P_s *cis*
Q_c	Compliance flow; $Q_c = Q - VA$ *cis*
t	Time *sec*
V_t	Total oil volume trapped between servovalve and actuator *in.*^{3}
β	Effective bulk modulus of oil = 150,000 psi
ζ_{lr}	Damping ratio of load resonance
ω_{lr}	Natural frequency of load resonance; $\omega_{lr} = \sqrt{(K_t + K_l)/M}$ or $\sqrt{G_t + G_l)/I}$ *rad/sec*
ω_{sv}	Natural frequency of servovalve *rad/sec*

For Linear Servos

A	Piston working area *in.*^{2}
F	Force output of piston; $F = A\Delta P$ *lb*
F_d	Force disturbance applied to load *lb*
K_a	Open loop stiffness of actuator and drive system; includes mechanical stiffness and oil spring $(4\beta A^2/V_t)$ *lb/in.*
K_{fv}	Velocity transducer gain $\frac{V}{in./sec}$
K_{fx}	Position transducer gain *V/in.*
K_l	Stiffness of load connection to ground (peculiar to pressure servos) *lb/in.*
K_s	Stiffness of actuator attach structure *lb/in.*

K_t	Net open loop drive stiffness; series combination of K_a and K_s *lb/in.*
K'_t	Series combination of K_a, K_s, K_l *lb/in.*
K_{vv}	Velocity gain of velocity loop sec^{-1}
K_{vx}	Velocity gain of position loop sec^{-1}
M	Mass of load *lb-sec²/in.*
V	Velocity of load relative to ground *in./sec*
V_f	Velocity following error *in./sec*
X	Position of load relative to ground *in.*
X_f	Position following error *in.*
X_u	Uncertainty in position due to anomalies in servovalve and drive system *in.*

For Rotary Servos

D	Motor displacement $in.^3/rev$
G_a	Open loop stiffness of motor and drive system; includes mechanical stiffness and oil spring ($\beta D^2/\pi^2 V_t$) *in.-lb./rad*
G_l	Stiffness of load connection to ground (peculiar to pressure servos) *in.-lb./rad*
G_s	Stiffness of motor attach structure *in.-lb./rad*
G_t	Net open loop drive stiffness; series combination of G_a and G_s *in.-lb./rad*
G'_t	Series combination of G_a, G_s, G_l *in.-lb./rad*
1	Load moment of inertia $in.\text{-}lb.\text{-}sec^2$
$K_{f\theta}$	Position transducer gain *V/rad*
$K_{f\Omega}$	Velocity transducer gain $\frac{V}{rad/sec}$
$K_{v\theta}$	Velocity gain of position loop sec^{-1}
$K_{v\Omega}$	Velocity gain of velocity loop sec^{-1}
T	Motor torque output; $T = (D/2\pi)\Delta P$ *in.-lb.*
T_d	Torque disturbance applied to load *in.-lb.*
θ	Position of load relative to ground *rad*
θ_f	Position following error *rad*

θ_u Uncertainty in position due to anomalies in servovalve and drive system *in.*

Ω Velocity of load relative to ground *rad/sec*

Ω_f Velocity following error *rad/sec*

(Courtesy of Moog, Inc., Buffalo, New York)

REFERENCES

1. John R. Luecke, *Fluid Power Communication Standards, 82.* National Fluid Power Association, Milwaukee, Wis., 1977.
2. W. J. Thayer, *Remote Control of Hydraulic Equipment.* From a paper presented at the Fluid Power Technology Update Conference held at the Colorado School of Mines. February 20–21, Moog Inc., East Aurora, N.Y., 1975.
3. T. P. Neal, *Performance Estimation for Electrohydraulic Control Systems.* From a paper presented at the National Conference on Fluid Power in Philadelphia, Pa., November 14, 1974. Moog Inc., East Aurora, N.Y.

7 Load Sensing Circuits

7.1. INTRODUCTION

Load sensing, as the name implies, relates to a circuit arranged to enhance sensitivity of control and provide economical operation in accordance with the immediate system power needs.

The purpose of sensing the load is to determine the precise power needed to attain the desired rotative speed or linear movement and add only that extra power needed for control purposes. The goal is to provide the desired power transmission function efficiently with safe operation and in some instances compactness. Also, any unnecessary additional power used in the control function, for whatever purpose, must be converted to heat if not into mechanical motion. Thus, the load sensing system reduces heat generation in the power transmission function as compared to equivalent nonload sensing circuits.

7.2. METER-IN LOAD SENSING DIRECTIONAL CONTROL VALVES

7.2.1. Basic Directional Control Valves

Directional, flow rate, and pressure control valves can be combined to precisely control the acceleration, rapid traverse, and deceleration of a work load.

The four-way directional control valve can be designed to incorporate load sensing pilot flow functions in this control activity in addition to directing the major power transmission flows within the circuit.

As an example, the directional control spool of Fig. 7.1(a) provides usual major directional flow functions. In addition, the spool directs flow from the spring chamber of the pressure control valve to tank in a series circuit in the neutral position. As the spool is operated, it blocks the flow to tank and connects a cylinder port to the pressure control valve spring chamber.

The pilot network shown in Fig. 7.1(a) serves to provide the needed transmission of a pressure signal reflecting the load needs which will establish desired pressure levels within the power transmission system in conjunction with the pressure control valves. A pilot control network is provided in the valve of Fig. 7.1(b) and 7.1(c) to accomplish a similar function.

The valve of Fig. 7.1(a) is thus provided with a pilot signal network that provides two functions. In neutral, the pressure control valve bias spring chamber is connected to tank through a pilot network. Shifting any directional valve will block this pilot circuit and direct pilot fluid from the working input cylinder port (B2) to the pressure control valve circuit and to the pressure control valve's control chamber or to the compensator of a variable delivery pump. Thus, the pressure will be additive to the basic bias spring and will reflect the cylinder load and system resistance.

The valves shown in Fig. 7.1(b) and (c) are designed with a *logic loop* [item 10, Fig. 7.1(b)] which will direct pilot flow fluid from the pressurized working cylinder port [item 9, Fig. 7.1(b)] to the control chamber of the pressure control valve or the compensator of a variable delivery pump. This will happen regardless of the valve in the circuit that is shifted. The flow of pilot oil is through control check valve 11 [Fig. 7.1(b)] to the pressure control mechanism. When the working spool or spools are returned to neutral, a small orifice bleeds the pilot fluid to tank so that a pressure valve will respond to the bias spring only or a variable delivery pump will relax to a low energy input setting.

The four-way directional control valve as described above will function as a compensated flow control valve. Movement of the directional control valve spool uncovers ports within the valve. The size of the opening is established by the operator or by a preset pilot actuator program. Note the ground angles on the spool of Fig. 7.1(a) and the notches (item 7) on the main spool of Fig. 7.1(b).

The flow through the resulting orifices created by the action of the valving members is maintained at a uniform pressure drop by means of the interrelationship between the four-way valve and an associated pressure control valve which is located between the flow source (pump, accumulator, etc.) and the input to the directional valve or the variable delivery pump control mechanism.

7.2.2. Pressure Level Control

Two types of pressure control valves are used to establish a constant pressure drop through the flow control orifice.

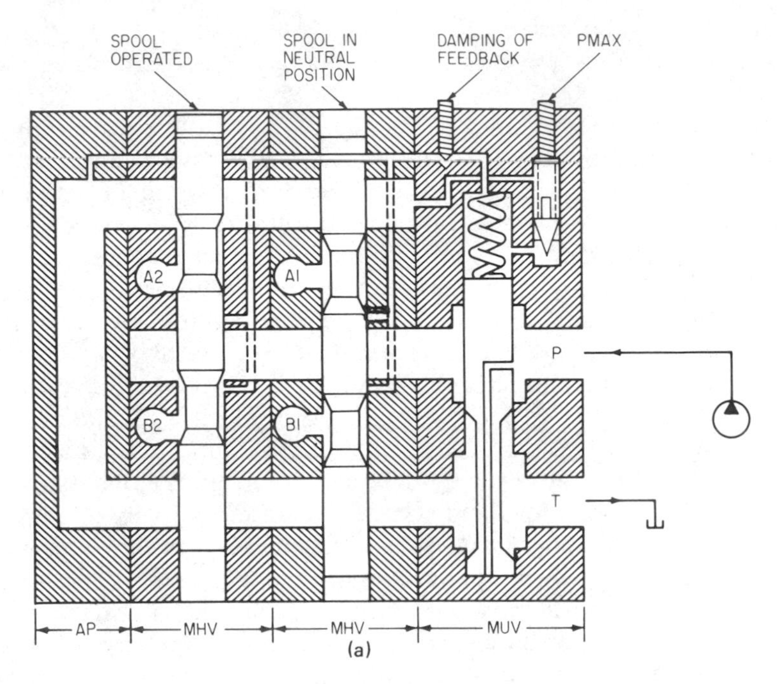

Figure 7.1 (a) Load sensing directional control valve assembly, fixed displacement pump. (Courtesy of HPI-Nichols, Sturtevant, Wisconsin.)

Normally-Closed Pressure Control Valves

One pressure control mechanism is structured with a normally-closed two-way valve serving as a pressure compensator which is maintained in the closed position by a bias spring. This bias spring is in control when a pilot control signal is not present. The bias spring in conjunction with the two-way valve structure establishes a minimum pressure difference between the pressure control outlet and the directional control valve inlet.

When a load is present at the cylinder port of the directional control valve, it is sensed in the pilot circuit [Fig. 7.1(a)] if the directional control valve element connects a cylinder port [B2, Fig. 7.1(a)] with the inlet of the directional control valve. The resulting pressurized pilot flow is additive to the pressure control valve bias spring. Thus the pressure upstream of the directional valve is a summation of the downstream load plus the pressure created by the valve bias spring. The net result is a pressure drop through the directional control valve passage equal to the value of the bias spring incorporated in the pressure control valve.

The normally-closed pressure control valve diverts excess fluid to a secondary conductor in which there is a lower pressure level. This occurs when

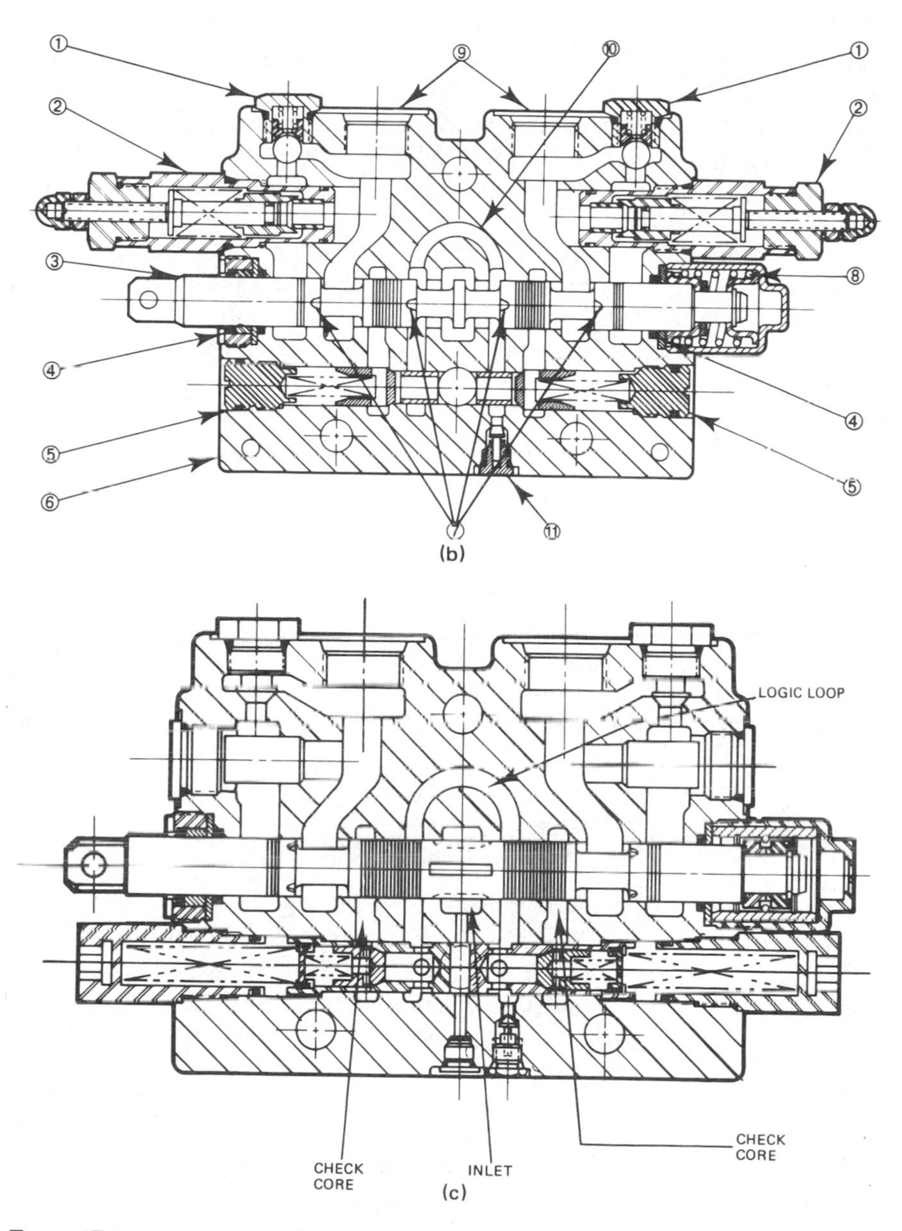

Figure 7.1 *(continued)* (b) Load sensing directional control valve with logic loop pilot sensing. 1: anticavitation check valve. 2: cylinder line relief valves. 3: directional spool. 4: externally replaceable seals. 5: load checks. 6: high-tensile gray iron casting. 7: metering notches. 8: centering spring. 9: cylinder ports. 10: looped core to supply signal through check. 11: to control circuit. (c) Load sensing directional control valve with inlet reducing valve-type pressure compensation for multiple valve operation. (Courtesy of Parker Hannifin Corporation, Mobile Hydraulic Division, Cleveland, Ohio.)

the flow available to the directional valve inlet is greater than the working flow passing through the directional valve or needed by the load(s). Often this secondary flow is directed to the reservoir with a minimum resistance encountered in the conductors. The normally-closed pressure control pilot circuit may include a small capacity relief valve which will permit a predetermined pressure and limit the pressure assist to the bias spring for a predetermined maximum value. Thus, this valve assembly can perform a relief valve, or maximum pressure control function. The result is the establishment of a maximum working pressure in the associated portion of the working circuit.

The pressure level established by the normally-closed pressure control is usually associated with a fixed displacement pump.

Variable Displacement Pump Circuit

A variable displacement pump used as the power and flow supply source can be assembled with an automatic control mechanism to adjust the output flow to match the demand created by the work cycle of the machine. One or more directional control valves can be supplied by a single variable displacement pump. Potential maximum flow needs are calculated as the machine circuit is developed. The variable displacement pump is sized so that it can supply adequate flow at maximum pressure and maximum flow levels of the machine operating cycle.

Normally-Open Pressure Control Valves

A type of normally-open compensator valve can be positioned in the supply line from the variable displacement pump outlet to the directional control valve inlet (Fig. 7.2). The flow through check valve 11 [Fig. 7.1(b)] would be shown in a circuit such as Fig. 7.2 in a similar manner and an orifice to tank would be shown beyond adjustable orifice *D* and the pressure additive chamber at spring chamber *B* to relax the signal pressure as the directional valve as indicated by orifice *A* is returned to neutral position. The series pilot circuit through the directional valves of Fig. 7.1(a) relaxes the pilot pressure signal in a similar manner to the fixed orifice to tank employed in the pilot circuitry of Fig. 7.1(b). A bias spring urges the control element in the compensator valve to the normally-open position. The signal to the compensator control element which will cause the valve to close is taken from the downstream conductor located between the compensator valve outlet and the directional control valve inlet (in Fig. 7.2 the directional control, similar to that in Fig. 7.1(a), is represented by adjustable orifice A). Thus the pressure difference between the outlet of this two-way valve and the inlet of the directional valve will be equal to the bias spring value when the directional control valve is in neutral or at rest position.

In the operational mode, the directional control valve will direct input flow to a cylinder port. A pressure signal is directed from this cylinder port

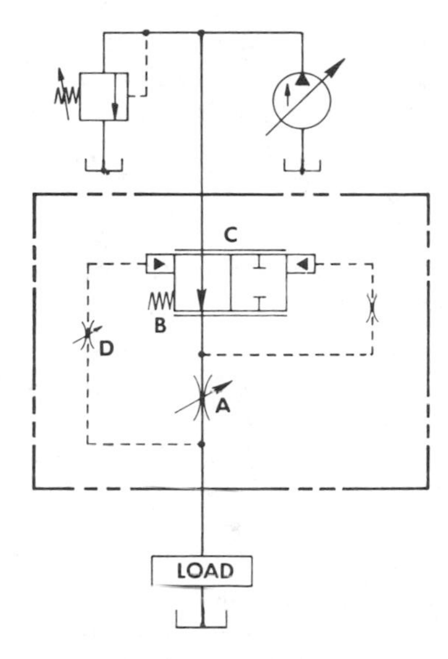

Figure 7.2 Load sensing directional control valve assembly, variable displacement pump. (Courtesy of HPI-Nichols, Sturtevant, Wisconsin.)

to the bias spring area of the normally-open compensator valve so that it can effectively add to the basic spring force. This pressure signal relates the load needs. Thus, as in the previous discussion of the normally-open versus the normally-closed compensator valve, the normally-open compensator is urged to a position permitting sufficient pressurized flow to maintain a pressure level ahead of the directional valve thereby creating a pressure drop across this major control orifice effectively equal to the value of the compensator valve bias spring. The pressure drop across the directional valve is constant from initial movement of the directional control valve spool until maximum flow rate is established appropriate to the supply capacity.

Summary

To develop a circuit which will provide the desired movement of a load at a predetermined degree of sensitivity it is necessary to provide the following circuit elements:

A source of pressurized fluid flow appropriate to the circuit needs.

A directional control mechanism to provide the required control of fluid to power the actuator and load.

Appropriate load holding valve or valves.

A signal from the load establishing the pressure and flow needs.

A pressure compensating means between the pump and the directional control valve to provide the appropriate pressure and flow across the directional valve thereby insuring a uniform rate of flow as set by the operator or a separate signal input control mechanism.

7.2.3. Multiple Valves

If a multiplicity of directional valves are incorporated into the circuit in a parallel, series, or series/parallel mode, it becomes necessary to modify the pilot circuit to insure a predetermined operational sequence.

The signal to the bias spring chamber of either the normally-open or normally-closed valve effectively adds to the value of the bias spring. When the directional control valve is in the neutral or rest position, it can be designed with a suitable valve section which will connect the pilot chamber adjacent to the pressure control valve section to the reservoir [Fig. 7.1(a)], or relax the pilot pressure from downstream of check 11 [Fig. 7.1(b) and (c)] through a suitable bleed orifice to tank with the spools in the center position. The cylinder port pilot connections are blocked and the control flow is relaxed to atmospheric pressure level. In this rest position the potential pressure level at the inlet to the directional control valve is established at the nominal pressure created by the compensator spool bias spring.

Parallel Connections

When two or more directional control valves are connected with their inlet connections in parallel, the pilot signal passages are connected in series [Fig. 7.1(a)]. Thus, the shifting of either directional valve spool will first close off the pilot tank passage so that the resulting pilot signal from the cylinder line [B2, Fig. 7.1(a)] is available to the bias spring chamber. The logic loop of Fig. 7.1(b) or (c) senses pilot pressure in like manner and directs it through check valve 11 to the pilot chamber of the pressure control device.

If two directional valves are shifted simultaneously, the pilot signal from the upstream valve will control the pressure level in the valve of Fig. 7.1(a). This is an acceptable work pattern for many machines.

Shuttle Valve Control of Load Sensing Line

The addition of a shuttle valve network in the pilot circuit for a valve such as in Fig. 7.1(a) will result in a control circuit which senses the highest work load pressure within the system and establishes this as the pressure available to each valve inlet. Check valve 11 [Fig. 7.1(b) or (c)] serves the same purpose, sensing the highest working pressure from the logic loops of the several valves and adds it to the bias spring value of the pressure control device.

A shuttle valve is structured with two inlet ports and one outlet port (Fig. 4.10). The inlet port with the highest pressure level will be connected to the outlet port and the other port will be blocked.

Thus, the pilot signal emanating from the two directional valves with parallel inlets will have the pilot signal connected to the inlets of the shuttle valve. The shuttle valve will direct the highest pressure signal to the load sensing control.

The pressure supplied to each valve will be equal to the highest load pressure that is sensed in the pilot signal circuit plus the pressure resulting from the bias spring.

Two directional control valves can be connected to a shuttle valve. An additional shuttle valve can be added to the pilot signal network with each additional directional control valve. If all valves have a parallel supply, they will operate together or in a random pattern. The outlet of the first shuttle is connected to one inlet of the second shuttle valve. The other inlet is connected to the pilot signal circuit of the third directional valve. An identical pattern is provided for the addition of each directional control valve section.

The shuttle valve pilot network will sense the highest pressure value within this parallel supply circuit and insure adequate pressure drop across the directional spools.

Certain circuits may require extremes of working pressure which minimizes the advantage of the uniform pressure drop across the directional control valves with the lesser load. This disadvantage can be eliminated by adding a compensator spool at the inlet of each directional control valve (Fig. 7.3) or at the inlet from supply through the load holding check to the cylinder ports [Fig. 7.1(c)]. The supply to each compensator spool is connected in parallel to the pressurized fluid source. A compensator spool is positioned in the conductor between the supply source and the line connecting the compensator spools serving each directional control valve. This valve could serve a relief function with a source employing a constant displacement pump. If so it would be a normally-closed device.

A variable displacement pump may be best served with a compensator modified to provide a basic minimum pressure. The compensator accepts the highest pilot signal from the shuttle valve network, or from downstream of checks 11 [Fig. 7.1(b)], and adds to the pressure established by the basic minimum pressure device (bias spring) to establish a pressure in the parallel supply conductor adequate for the highest pressure needed by the highest load that was sensed in the pilot circuit.

The pressure sensed at the downstream load of any parallel supply directional control valve is first directed to the compensator immediately upstream of this directional control valve spool. Thus the pressure drop across the orifice created by the flow directing mechanism (notches in spool) is constant so long as an adequate supply is insured. This pressure is independent of all other branches within the circuit. This pilot signal is also directed through the shuttle network (Fig. 7.3). If it is the highest pressure sensed in the control network, it will establish the maximum pressure level to be maintained

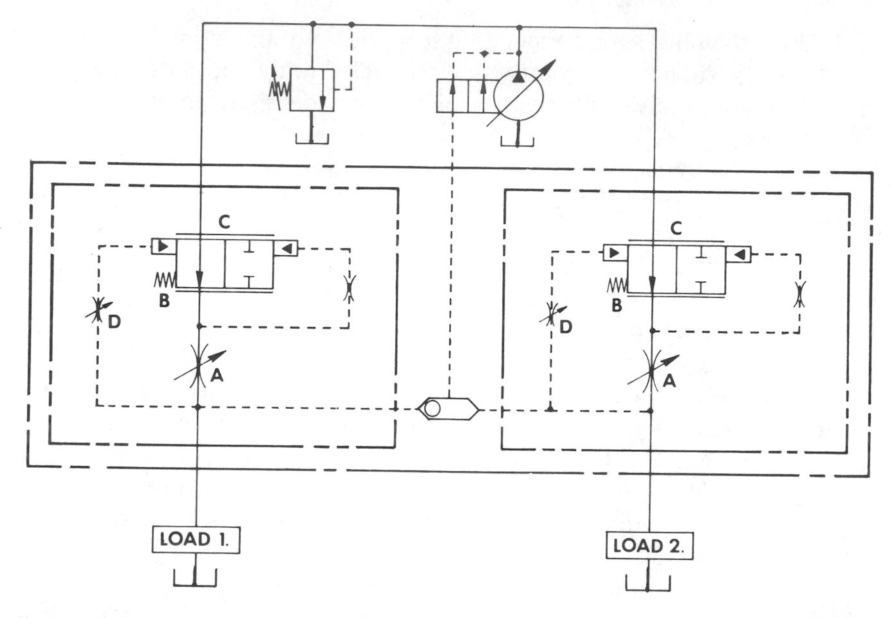

Figure 7.3 Load sensing variable pump control. (Courtesy of HPI-Nichols, Sturtevant, Wisconsin.)

by the supply pump. If it is less than another branch in the circuit, it will be blocked by the appropriate shuttle valve automatically.

Pressure Level Summary

The pressure levels within the various parts of the hydraulic circuit will be established by the following criteria:

Rest Condition The supply pump, if it is of fixed displacement, will deliver fluid to the parallel directional control valve inlets and the inlet of a normally-closed compensator valve.

The pressure is established in the rest position by the bias spring in this normally-closed compensator valve whose discharge is usually connected with little restriction to the reservoir. The pilot signal to the compensator valve is connected to the upstream side of the valve.

If the supply pump is of a variable displacement design, it will deliver fluid to the inlet of a normally-open compensator whose pilot signal is connected to the downstream side of the valve so that it will close at a pressure value established between the outlet of the compensator and the inlet of the parallel supply network to the directional control valves. In the rest position, this will be at the value created by the bias spring in the normally-open compensator. The pump can be supplied with a bias spring for a predetermined minimum pressure which will be the level at rest. As a signal is applied it is added to the minimum.

Load Sensing As the neutral status of the directional control valve bank is altered (by shifting one or more of the spools), a resulting block of the pilot tank passage will occur [Fig. 7.1(a)] and a load at the valve cylinder port will be sensed. This load will be additive to the compensator spool bias spring. The signal will be sensed through check(s) 11 [Fig. 7.1(b) and (c)].

Priority The upstream load will establish the pressure level if a single compensator is provided to establish a pressure level at the inlet of the directional valves.

The introduction of a normally-open compensator at the inlet of each directional valve spool will permit random operation of the directional valves with the control of the compensator at the outlet of the supply pump receiving the highest load signal via the shuttle valve network (Fig. 7.3) or through check valve 11 [Fig. 7.1(c)]. Thus the pump will deliver pressurized fluid at the maximum pressure needed to the supply network. Each branch compensator will separately establish the pressure level at the directional control valve inlet at the specific load value sensed at the cylinder outlet port of the directional control valves.

Pump Control The constant displacement pump will deliver fluid through the normally-closed compensator to the reservoir at a low pressure in the rest position of the circuit. As a directional control valve is actuated, some portion of the pump delivery will be directed through to the actuator. The remaining fluid will be directed to the reservoir at working pressure.

A variable displacement pump control mechanism can be structured to maintain maximum pressure at the inlet of the branch circuits. This can result in significant energy losses when the pump compensates or comes to neutral during standby services pumping only sufficient fluid to make up for internal leakages at this relatively high pressure. To function at maximum energy conservation levels, the pump is structured with a control that will reduce pressure to a predetermined minimum (often 200 psi) during the machine rest position. As the sensing network dictates need for higher flows and pressures, the relating signal pressure is directed to the pump control. The pump then operates at the appropriate flow and pressure level to satisfy maximum circuit needs at the highest pressure required by circuit loading.

7.2.4. Summary

Load sensing circuits have many excellent characteristics:

Intelligent system allows horsepower management because of load matching.

Constant gain metering allows good local proportional control, including multifunctional operation.

No metering with engine speed.

Pump loading during cycle related directly to operational loads only.

Possible reduction of pump displacement.

Generally lower operating temperatures.

Less standby horsepower.

Minimum throttling.

Enhanced safety.

Permits customer-oriented customized applications.

7.3. INTEGRATED METER-IN AND METER-OUT LOAD SENSING CONTROL SYSTEMS

Excavators, cranes, loaders, and similar machinery can benefit from high pressure systems because of the reduction in physical size of the hydraulic components [1]. The use of high pressure (3500–5000 psi) offers other significant benefits. With these benefits come some system requirements that may be satisfied in a somewhat different manner than with systems operating at medium or low pressures. The following four factors should be considered:

1. Hydraulic pilot operation may best replace direct manual operation of major valves.
2. A variable pump in an operational open circuit provides an efficient source of pressurized fluid.
3. Load sensing systems may offer best sensitivity and most efficient operation.
4. The capability of bringing a load to a controlled stop in case of line rupture is required by legislation in certain areas.

Other considerations are: (1) low leakage on cylinder ports is desired, (2) spool bind must be minimized, and (3) cavitation protection is essential.

Long valve spools in closely-fitted bores may be best suited for medium pressures. Poppet valves with integral metering facilities offer a useful alternate to long spool valves in high-pressure control devices.

A system which incorporates all of the factors necessary for efficient hydraulic power transmission at high pressures is illustrated in Fig. 7.4.

7.3.1. Supply Pump

The variable displacement pump of Fig. 7.4 incorporates a full flow supercharge pump (1). An external filter (2) cleans all fluid before it enters the system. Excess charge pump flow is available to backpressure the return line

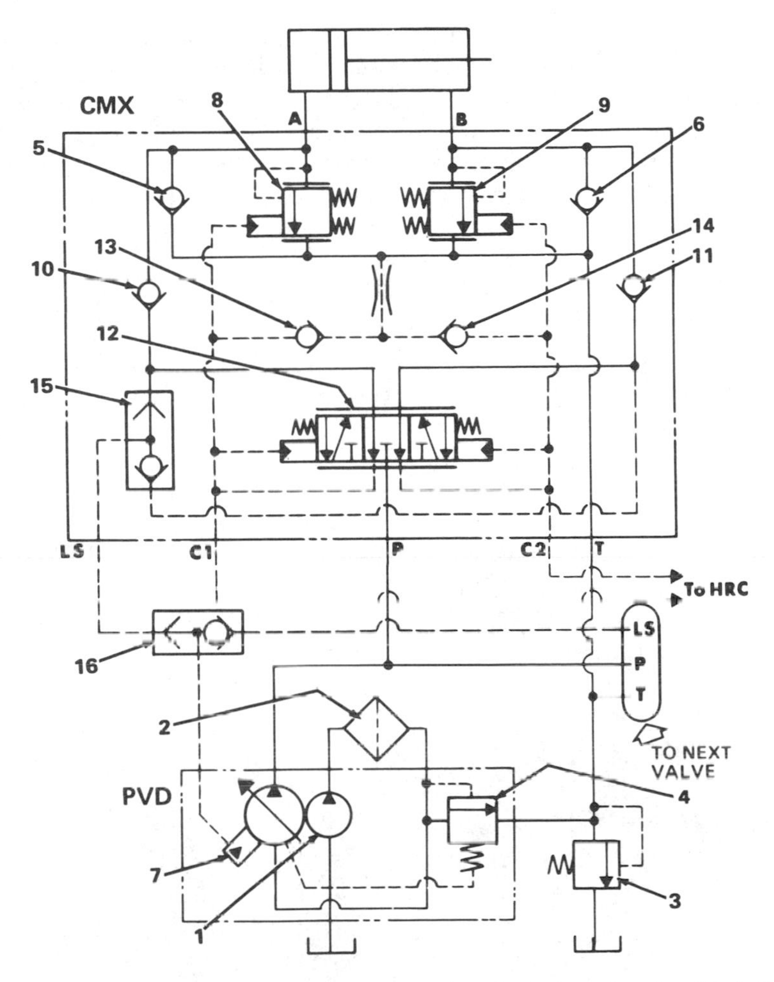

Figure 7.4 Integrated meter-in/meter-out load sensing circuit with full supercharge variable displacement pump. (Courtesy of Sperry Vickers, Troy, Michigan.)

as established by the pressure level control valve (3). The excess full flow charge pump delivery passes through the pressure level control valve (4) to a point upstream of the return pressure level control valve (3). The drain from the spring chamber of valve (4) is directed to the pump housing which is held at a relatively low pressure so that flow variables will not adversely affect the pressure level established by valve (4). The backpressure resulting from the variable restriction of valve (3) permits the anticavitation check valves (5 and 6) to function even when there is insufficient fluid returning from the rod end of a cylinder and pump flow is not being used because the load is overrunning. The backpressure also permits a quiescent flow in the pilots for the purpose of warming and purging. The load sensing compensator (7) keeps the pressure 200 psi plus line losses above the load pressure whenever the meter-in element is operating and flow is within capacity of the system. Typically, the pump would also incorporate input torque and pressure override controls.

7.3.2. Control Valve Assembly

All of the elements enclosed in the centerline enclosure are built into a single housing (Fig. 7.5). This valve is designed to be manifolded to an actuator (linear or rotary) (Fig. 7.6). Facility for a 6000 psi four-bolt flange are provided at the input pressure port. A low pressure four-bolt flange connection is provided at the the return to tank port. Pilot lines from the remote hydraulic joystick control (Fig. 7.7) or single function remote control (Fig. 7.8) are connected to the valve of Fig. 7.5 with tube or hose. Since the valve is mounted on an actuator, there is minimal pressure drop between the valve and a cylinder or motor. Cavitation protection is thus greatly improved.

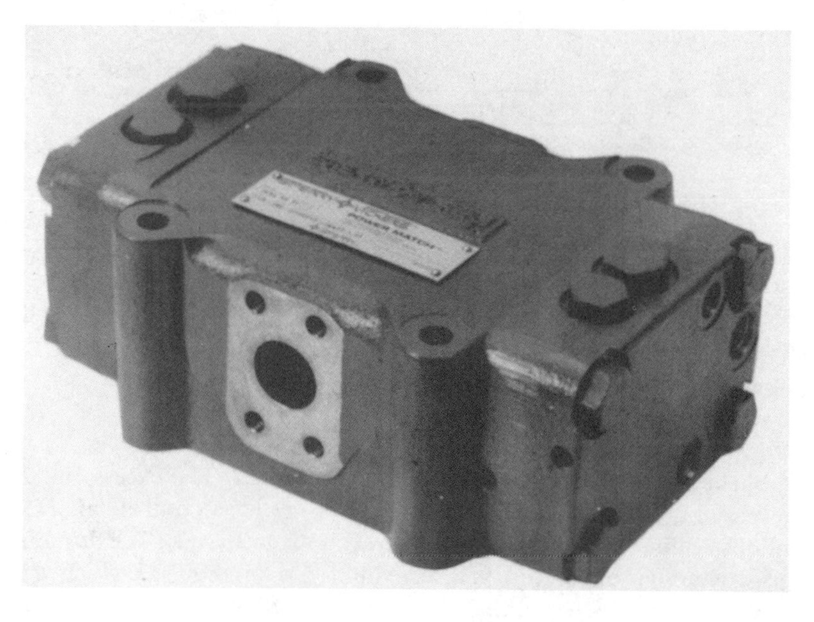

Figure 7.5 Integrated meter-in/meter-out load sensing valve. (Courtesy of Sperry Vickers, Troy, Michigan.)

Figure 7.6 Valve for direct mounting to actuator. (Courtesy of Sperry Vickers, Troy, Michigan.)

7.3.3. Safety Considerations

A review of the schematic reveals that any of the external connections to the valve, except "A" and "B" ports which are manifolded connections, can be ruptured without loss of ability to bring a load to a controlled stop. The same elements within the valve are used for both normal operation and line rupture protection. Thus, the problems associated with transferring the blocking function from one element to another are avoided. Relief protection is continuously provided by the meter-out poppet valves (8 and 9 of Fig. 7.4). Compressibility of fluid in interconnecting lines does not result in objectionable load movement.

Cylinder port lines typically experience higher pressure levels than do supply lines. Even in the case of high-response relief valves without overshoot, port relief settings are higher than supply pressure because of the need to provide a margin to decelerate a load and to eliminate interaction. By actuator mounting of a valve, interconnecting lines and hoses are not subjected to these

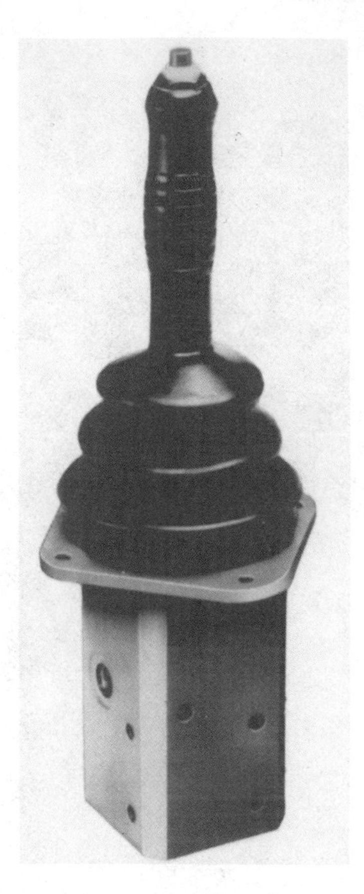

Figure 7.7 Joystick control for remote pilot actuator. (Courtesy of Sperry Vickers, Troy, Michigan.)

high pressures. The number of pressure cycles are also reduced, since the pressures encountered in decelerating a load to a stop are not experienced beyond the cylinder port.

7.3.4. Operation of Valve Elements

Operation will be described by first considering the meter-out function and then the meter-in function. Meter-out is the control of exhaust fluid from a motor or cylinder to a reservoir. Meter-in is the control fluid from the pump to a motor or cylinder.

Meter-out Element

The meter-out element is shown in Figs. 7.9 (and items 8 and 9 of Fig. 7.4). It incorporates a stem which is positioned by the force balance which exists between a pilot operated piston and spring. The stem and meter-out poppet serve as a simple bleed servo. When the stem is moved to the left by command pilot pressure at "C1" acting on the pilot piston opposed by the spring, fluid is allowed to drain from behind the meter-out poppet. Pressure acting on the left end of the poppet is reduced, since the chamber is supplied by flow through an orifice from cylinder port "A". The poppet will be moved to the left by

Figure 7.8 Individual controls for remote pilot actuation. (Courtesy of Sperry, Vickers, Troy, Michigan.)

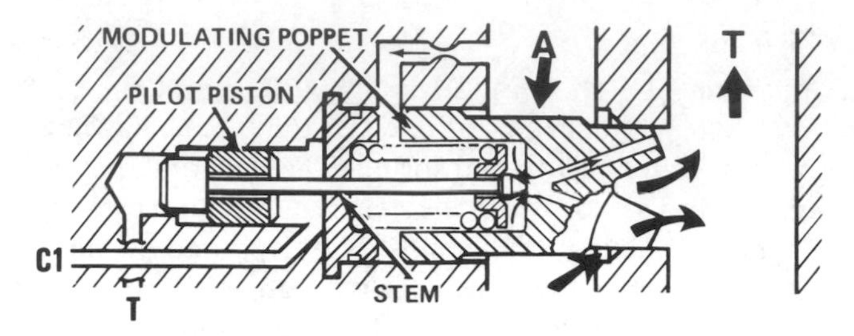

Figure 7.9 Meter-out control assembly. (Courtesy of Sperry Vickers, Troy, Michigan.)

cylinder port "A" pressure acting on the differential area. As the poppet approaches the stem, the restriction between the left end of the poppet and tank will increase, and pressure in the chamber will rise. The poppet will come to rest when pressure in the chamber rises sufficiently to balance cyinder port "A" pressure acting on the differential area. Because of the magnitude of the forces available to position the meter-out poppet, a very close relationship exists between the stem position and the meter-out poppet position. Flow forces acting on the meter-out poppet have little effect. Also, frictional forces acting on the meter-out poppet have little effect. Essentially, the meter-out element serves as a variable orifice between one of the actuator ports and tank, with the degree of opening controlled by the magnitude of the command pilot pressure at "C1".

The *bar* is a measure of pressure which equals 14.5 psi. Many high pressure systems are calibrated and instrumented in metric units. Obviously, descriptive literature is in metric units. Metric units are used in describing the following operational characteristics.

The relationship between flow through the meter-out element and command pilot pressure for an overrunning load pressure of 138 bar is shown in Fig. 7.10. This test was run by connecting cylinder port "A" of a valve to a source which was maintained at 138 bar. Command pressure was slowly increased from zero to greater than 10 bar, then slowly decreased. Flow was continuously plotted as a function of command pressure. Friction will reduce the accuracy with which flow will follow command pressure, and will cause a separation (hysteresis) between the lines for increasing and decreasing pressure. Because the stem and poppet comprise a simple servo system, only that friction which acts upon the pilot piston and the pilot stem causes hysteresis. There is, therefore, minimal hysteresis associated with the valve element. As a result of minimizing hysteresis and the effects of flow forces, the ability to control the lowering of loads is greatly improved over direct pilot control of spools.

Relief Valve

A second mode of control is provided to enable the meter-out element to also provide relief protection. Figure 7.11 shows the relief valve configuration. The pilot spool diameter is slightly larger than that of the balance piston. On a steady state basis, the balance piston is forced to the left by a force which is equal to the product of cylinder port pressure and the differential area between the pilot spool and the balance piston. When cylinder port pressure exceeds a value where the force overcomes the spring, which acts as bias for the balance piston, the pilot spool opens allowing fluid to exhaust from behind the meter-out poppet. This results in a decrease in the pressure acting on the left end of the meter-out poppet. Cylinder port pressure acting on the area differential will force the meter-out poppet to the left and will allow fluid to exhaust from the cylinder port to the tank. This serves as a means of limiting the pressure at cylinder port "A." Override of the relief valve for gradual changes in cylinder port flow is shown in Fig. 7.12. There is a minimal override, which permits actuators to be subjected to lower pressures when relief valve flow is high.

There is an accumulator chamber which can communicate with the cylinder port through an orifice. As pressure changes in the cylinder port,

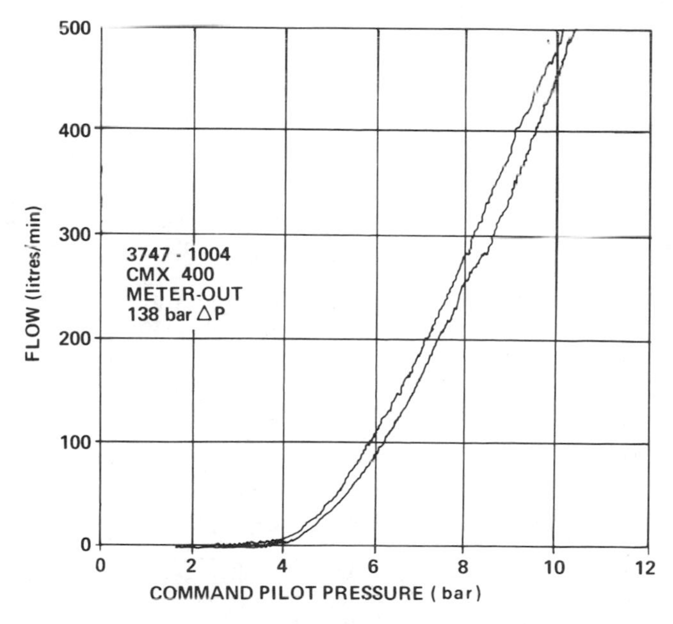

Figure 7.10 Flow through meter-out element command pressure. (Courtesy of Sperry Vickers, Troy, Michigan.)

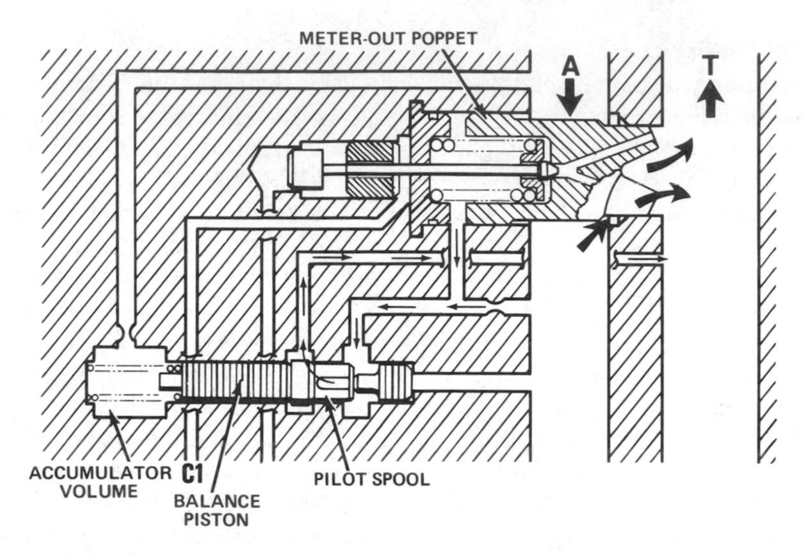

Figure 7.11 Port relief valve assembly. (Courtesy of Sperry Vickers, Troy, Michigan.)

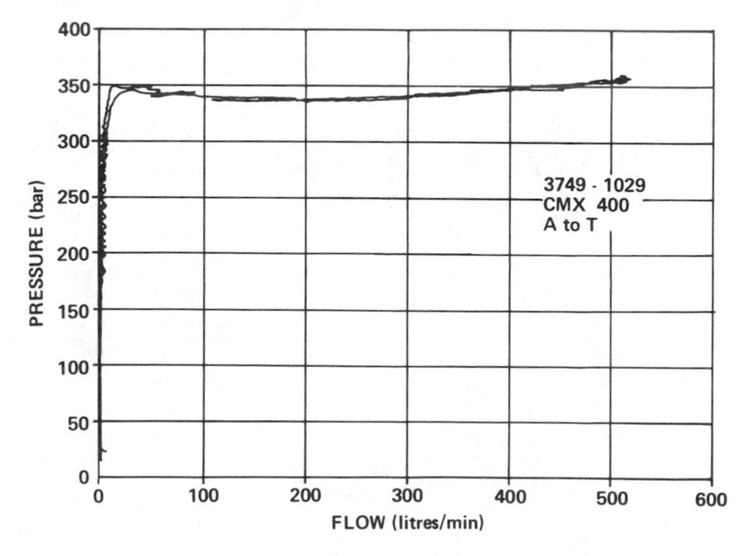

Figure 7.12 Relief valve override curve. (Courtesy of Sperry Vickers, Troy, Michigan.)

the accumulator pressure will also change, but will lag behind that of the cylinder port. If the pressure in the cylinder ports tends to rise at a rate exceeding roughly 4000 bar per second, the reduced pressure acting on the balance piston will allow the pilot spool to open even though the cylinder port pressure may be considerably below its normal maximum pressure setting. The net effect of this is the relief valve not only controls maximum cylinder port pressure, but it also limits the rate of pressure rise. Dynamic response of a relief valve is shown in Fig. 7.13. The test run is started by discharging 400 l/min. into a circuit and suddenly blocking a bypass valve. The cylinder port relief valve must react to limit pressure. It can be seen that the rate of pressure rise as pressure approaches the maximum setting is below the initial rise rate of almost 15,000 bar/sec. Even with such a high initial rate, overshoot above the maximum setting is eliminated. Overshoots of 30% or more are common in conventional relief valves.

When translated to the effect on machine operation, the pressure limiting feature limits the rate of change of acceleration, or jerk. The elimination of overshoot limits the maximum force that can be exerted by a cylinder or the maximum torque that can be exerted by a motor. Not only is machine operation smoother, but also reliability is improved because of reduced fatigue of mechanical and hydraulic components.

Meter-in Element

Figure 7.14 shows the meter-in element in the center position. Command pressure can be applied at either "C1" or "C2", depending upon the desired direction. Consider the case when pilot pressure is applied at "C1". When the

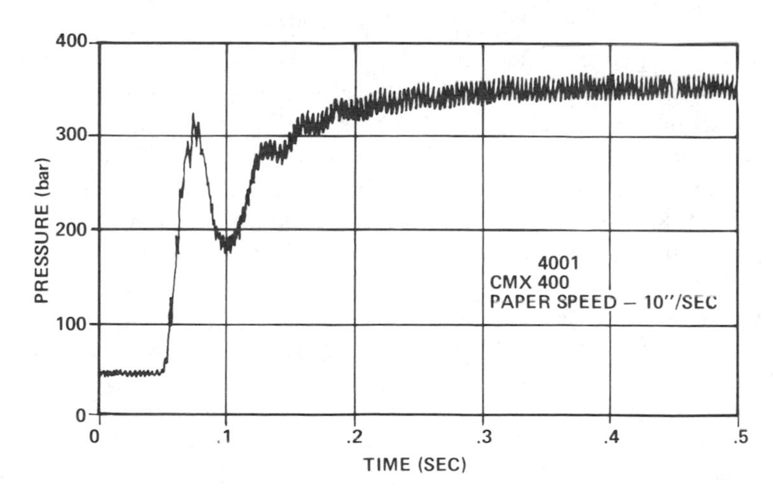

Figure 7.13 Dynamic response of relief valve. (Courtesy of Sperry Vickers, Troy, Michigan.)

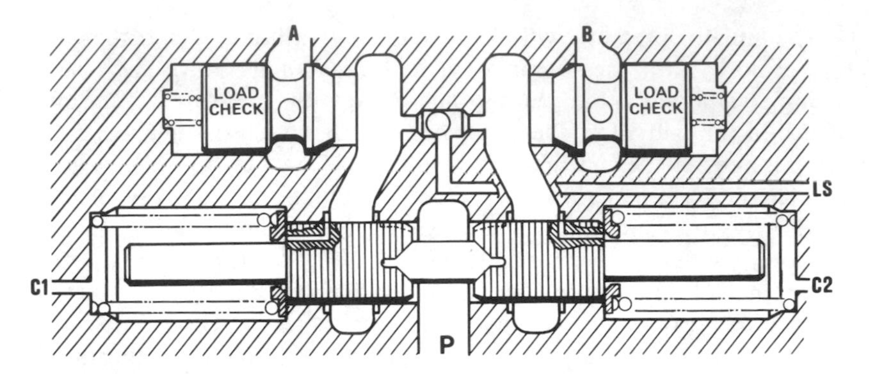

Figure 7.14 Meter-in element in center position. (Courtesy of Sperry Vickers, Troy, Michigan.)

command pressure reaches a level so that the force acting on the spool overcomes the initial spring preload, the spool will move to the right and block the drain for the chamber between the meter-in spool and cylinder port "B" load check valve. Continued increase in the pressure at "C1" will result in opening the area between "P" and cylinder port "B", and flow will occur. The shuttle valve will shift to the left and load pressure will be admitted to the port "LS." In a load sensing system, load pressure is fed back from "LS" to the pump. The pump displacement will be controlled such that the output flow will result in the pressure at "P" being higher than the pressure at "LS" by approximately 14 bar. Precise metering of flow to the load can be achieved by varying the pilot command pressure "C1". Differential pressure across the meter-in spool is maintained at a fixed value by the pump as long as the demand is within the maximum limits of pressure, input torque, and pump flow. When the command pressure at "C1" is diminished, the meter-in spool will center and the load sensing signal will be allowed to decay by flow through the small drilling in the meter-in spool at its centered position. By picking up the load signal in such a manner, *a continuous bleed is avoided*. Also, pressure is present between the meter-in spool and the load check only when a load is being driven. There is no need to resort to complex spool and casting configurations to prevent pressure arising from overrunning loads from interfering with the load sensing signal.

Figure 7.15 shows the relationship between meter-in flow and command pressure at a meter-in differential pressure of 345 bar. It can be seen that frictional effects are minimal, and thus it is possible to get smooth control of flow to an actuator.

Multiple Valve Operation

When two valves are supplied by a single pump and both valves are operated simultaneously, it is possible for one valve to have an excessively high pressure drop across the meter-in element because of the difference in the loads. Shuttle valve 16 (Fig. 7.4) senses the highest load pressure. This is signalled to the load sensing pump. For example, it is possible for the pressure drop to increase from a normal load sensing differential pressure of 14 bar to the maximum supply pressure of 350 bar. Unless some form of compensation is used, the flow rate going to the lower pressure actuator might increase by a factor of 5 as a second function is brought into service. Manufacturers have typically used a separate full flow inlet pressure compensator spool to eliminate this undesirable result.

With the integrated meter-in and meter-out load sensing system, the

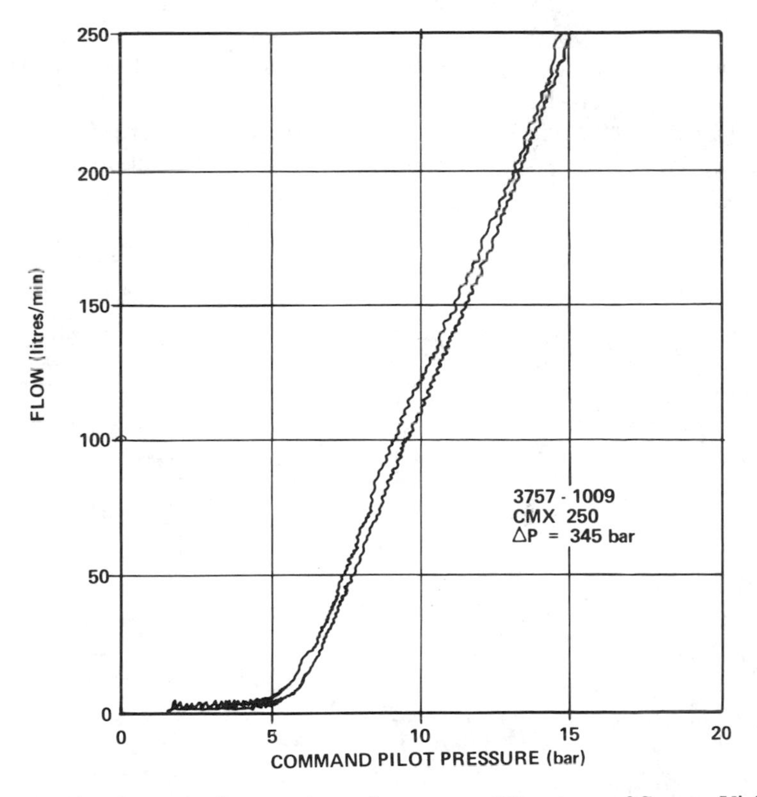

Figure 7.15 Meter-in flow command pressure. (Courtesy of Sperry Vickers, Troy, Michigan.)

metering area gain and spring rates have been selected so that inlet flow force compensation occurs. Fig. 7.16 shows meter-in flow versus differential pressure for three values of command pilot pressure. For a command pressure 15 bar, the pressure drop across the meter-in element increases from 14 bar to 350 bar while flow increases by only about 20% at the maximum flow point. If flow force compensation were not incorporated, the increase would be about 400%. The compensation makes it possible for an operator to simultaneously control two functions with minimal interference from one function on another.

Meter-in and Meter-out Phasing

Normally the meter-out element will begin to open at a pilot pressure of 3.5 bar and be fully open at 14 bar. The meter-in element would begin to open at a pilot pressure of 10 bar and be fully open at 21 bar. The phasing of the meter-out and meter-in elements are plotted against the hydraulic remote control input angle in Fig. 7.17. It will permit an operator to control an overrunning load over a considerable speed range without utilizing meter-in flow. He would also have a significant range of metering for a load which must be powered by pump flow.

If a load is overrunning and it is not necessary to power the load at high speed, the hydraulic remote control characteristics can be selected so that the

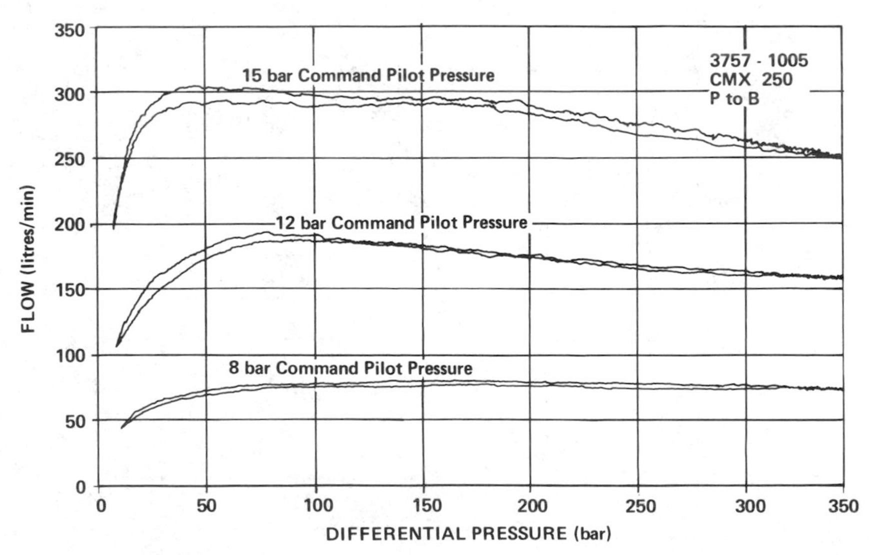

Figure 7.16 Meter-in flow differential pressure. (Courtesy of Sperry Vickers, Troy, Michigan.)

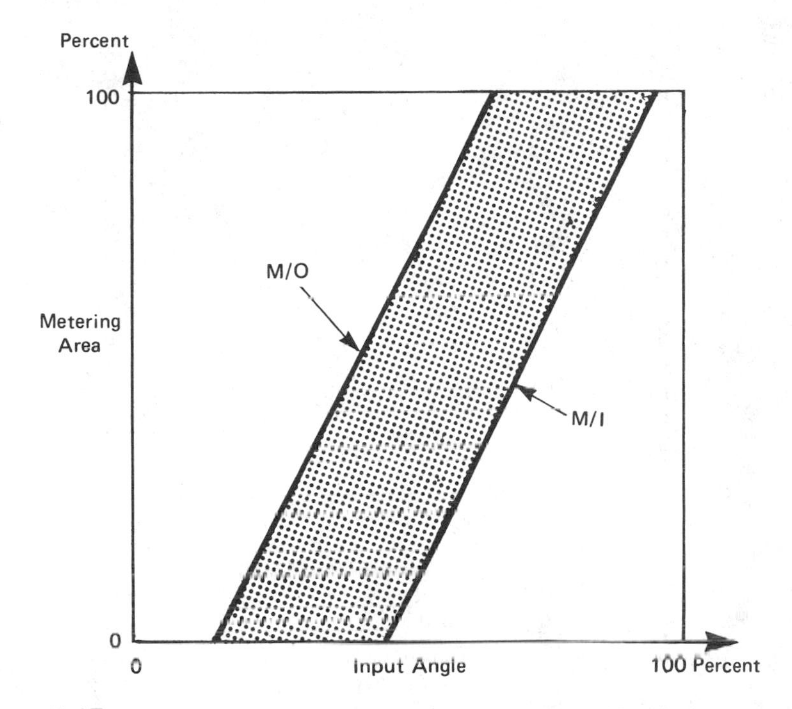

Figure 7.17 Metering arrangement for general type of load. (Courtesy of Sperry Vickers, Troy, Michigan.)

meter-in element will open only a slight amount. The resulting phasing is shown in Fig. 7.18. It might be desirable on an application such as an excavator where it is not necessary to power the boom downward at high speed. Such phasing would keep an operator from inadvertently robbing flow from another function. When lowering without pump flow, fluid to fill the supply side of the cylinder would be provided through anticavitation check valves from the tank port.

7.3.5. Operation of the Complete Valve

The operation of the valve elements has been described. The relationship of the elements in a complete valve will now be considered.

Neutral Condition

Figure 7.19 shows a functional schematic for a valve in a standby or neutral condition. Meter-out elements are blocked (Fig. 7.4, meter-out valves 8 and 9, anticavitation checks 5 and 6, load checks 10 and 11), since pilot pressure is not being applied and the cylinder ports are below relief valve settings.

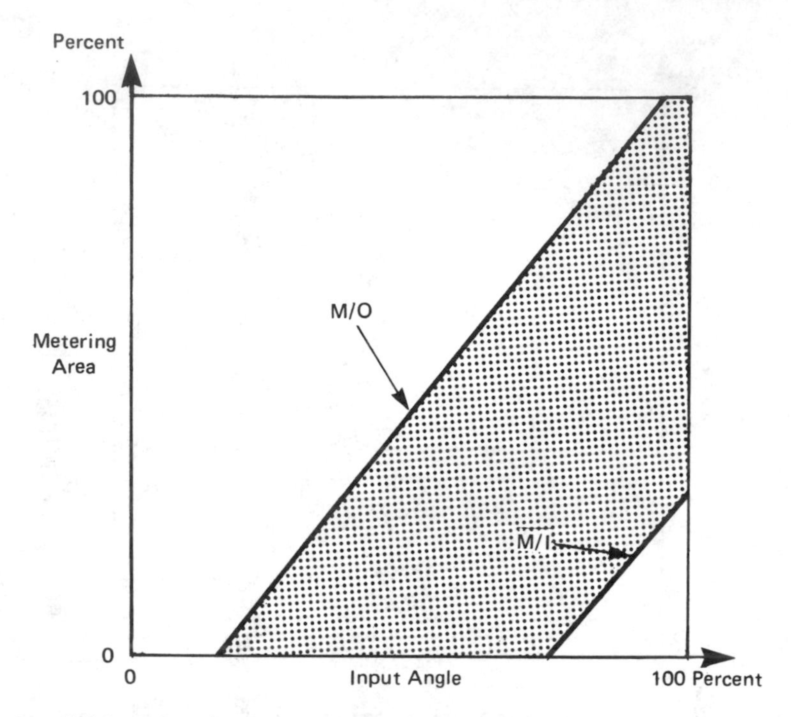

Figure 7.18 Metering arrangement for primary overrunning loads. (Courtesy of Sperry Vickers, Troy, Michigan.)

Cylinder port relief protection is available if needed (Fig. 7.4, relief valve section of meter-out valves 8 and 9). Anticavitation check valves (Fig. 7.4, check valves 5 and 6) and load check valves (Fig. 7.4, check valves 10 and 11) are closed. The meter-in spool (Fig. 7.4, valve 12) is centered and the load sensing signal has decayed through either "C1" or "C2". With the exception of the relief valve balance piston and the meter-out system, no fixed clearance sealing elements communicate with the cylinder ports. Instead, poppet type seals are used. Leakage is considerably less than that of a conventional sliding spool valve.

Lowering a Load

Figure 7.20 represents the condition when pilot pressure has been applied at "C1" sufficient to open the meter-out element of port "A". The pilot piston on the stem of the meter-out valve (item 8 in Fig. 7.4) acts as a check valve as it contacts the head of the stem assembly (This is check valve 13, Fig. 7.4). Preloaded springs keep the meter-in spool centered (item 12, Fig. 7.4). The load forces the cylinder to the left. The anticavitation check valve at port "B"

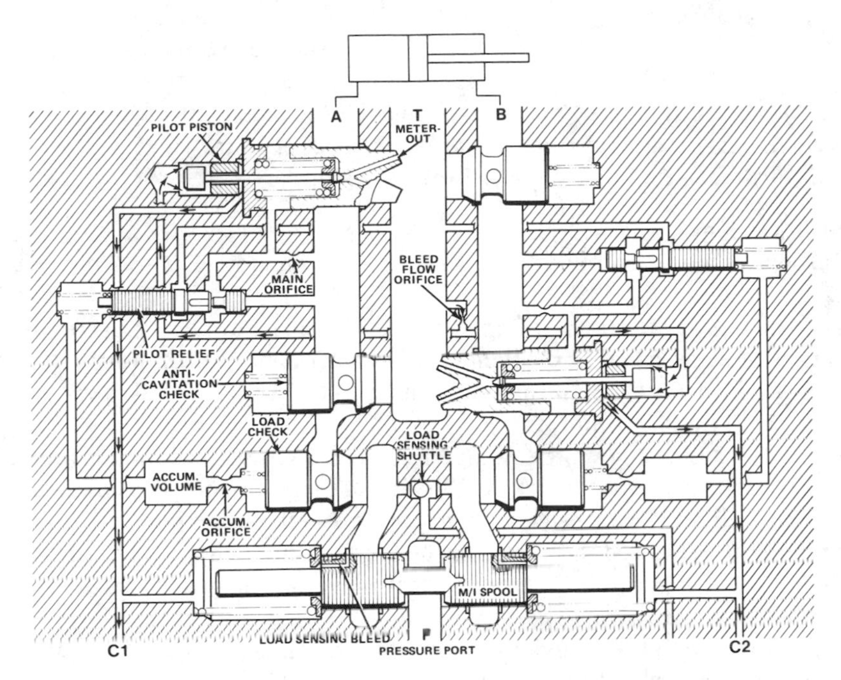

Figure 7.19 Neutral conditions. (Courtesy of Sperry Vickers, Troy, Michigan.)

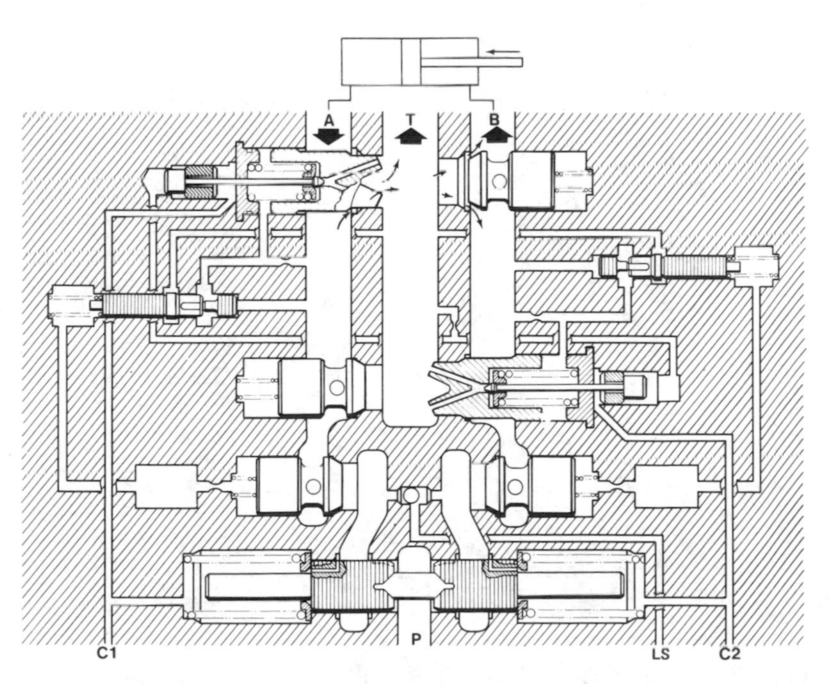

Figure 7.20 Lowering a load. (Courtesy of Sperry Vickers, Troy, Michigan.)

(Check valve 6, Fig. 7.4) opens to permit the rod side of the cylinder to be filled. Excess return oil flows to tank over backpressure valve 3, Fig. 7.4. Pump flow can be utilized for another function.

Driving a Load

The control valve assembly functions when driving a load as shown in Fig. 7.21.

Command pressure acts on the meter-out element of port "A" permitting flow from the head end of the cylinder to tank. Command pressure also opens the meter-in element admitting flow from port "P" past the load check valve (item 11, Fig. 7.4), to port "B."

The load sensing shuttle ball (item 15, Fig. 7.4) shifts to the left, feeding load pressure to the port marked "LS." From there, the signal is fed back to a suitable variable displacement open loop hydraulic circuit pump. The compression flow required to pressurize the load sensing line will not cause cylinder movement because of the load check valve (signal is taken ahead of the closed load check valve). The pressure at "P" must be higher than load pressure. Flow rate to the cylinder is controlled by manually controlling an hydraulic remote control to select pressure level at "C1."

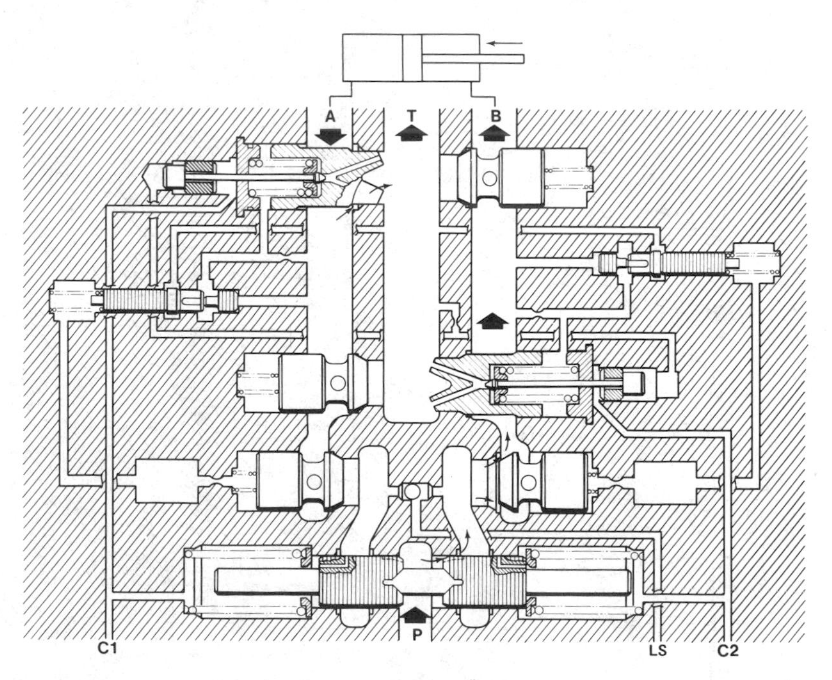

Figure 7.21 Driving a load. (Courtesy of Sperry Vickers, Troy, Michigan.)

7.3.6. Summary

High-pressure load sensing valves as shown in the structure of Fig. 7.4 have been developed primarily for earthmoving and construction equipment. They are actuator mounted and operated by remotely commanded pilot pressure. Multiple valves can be parallel connected to the outlet of a suitable load sensing variable open hydraulic circuit pump.

The valves have meter-out elements which also serve as pressure rate limiting valves providing a relief function. Phasing of meter-in element with respect to the meter-out elements can be preselected to provide an optimum match for various load types. Interaction between simultaneously operated valves is minimized by meter-in pressure compensation. Cylinder port leakage is controlled by wide usage of poppet type seals. Actuator mounting permits load control even if a line is ruptured. It also improves cavitation protection, eliminates problems associated with multiple elements used for line rupture protection, and reduces both the magnitude and number of pressure cycles that interconnecting lines must withstand.

REFERENCE

1. Robert H. Breeden, *Development of a High Pressure Load Sensing Mobile Valve*. From SAE Technical Paper 810697 presented at the Earthmoving Industry Conference, sponsored by Society of Automotive Engineers, Inc., Peoria, Illinois. Sperry Vickers Div., Sperry Corp., Troy, Michigan, April 6–8, 1981.

Index